Fundamentals of Multicore Software Development

Chapman & Hall/CRC
Computational Science Series

SERIES EDITOR

Horst Simon
Deputy Director
Lawrence Berkeley National Laboratory
Berkeley, California, U.S.A.

AIMS AND SCOPE

This series aims to capture new developments and applications in the field of computational science through the publication of a broad range of textbooks, reference works, and handbooks. Books in this series will provide introductory as well as advanced material on mathematical, statistical, and computational methods and techniques, and will present researchers with the latest theories and experimentation. The scope of the series includes, but is not limited to, titles in the areas of scientific computing, parallel and distributed computing, high performance computing, grid computing, cluster computing, heterogeneous computing, quantum computing, and their applications in scientific disciplines such as astrophysics, aeronautics, biology, chemistry, climate modeling, combustion, cosmology, earthquake prediction, imaging, materials, neuroscience, oil exploration, and weather forecasting.

PUBLISHED TITLES

PETASCALE COMPUTING: ALGORITHMS AND APPLICATIONS
Edited by David A. Bader

PROCESS ALGEBRA FOR PARALLEL AND DISTRIBUTED PROCESSING
Edited by Michael Alexander and William Gardner

GRID COMPUTING: TECHNIQUES AND APPLICATIONS
Barry Wilkinson

INTRODUCTION TO CONCURRENCY IN PROGRAMMING LANGUAGES
Matthew J. Sottile, Timothy G. Mattson, and Craig E Rasmussen

INTRODUCTION TO SCHEDULING
Yves Robert and Frédéric Vivien

SCIENTIFIC DATA MANAGEMENT: CHALLENGES, TECHNOLOGY, AND DEPLOYMENT
Edited by Arie Shoshani and Doron Rotem

INTRODUCTION TO THE SIMULATION OF DYNAMICS USING SIMULINK®
Michael A. Gray

INTRODUCTION TO HIGH PERFORMANCE COMPUTING FOR SCIENTISTS
AND ENGINEERS, Georg Hager and Gerhard Wellein

PERFORMANCE TUNING OF SCIENTIFIC APPLICATIONS, Edited by David Bailey,
Robert Lucas, and Samuel Williams

HIGH PERFORMANCE COMPUTING: PROGRAMMING AND APPLICATIONS
John Levesque with Gene Wagenbreth

PEER-TO-PEER COMPUTING: APPLICATIONS, ARCHITECTURE, PROTOCOLS, AND CHALLENGES
Yu-Kwong Ricky Kwok

FUNDAMENTALS OF MULTICORE SOFTWARE DEVELOPMENT
Victor Pankratius, Ali-Reza Adl-Tabatabai, and Walter Tichy

Fundamentals of Multicore Software Development

Edited by
Victor Pankratius
Ali-Reza Adl-Tabatabai
Walter Tichy

CRC Press
Taylor & Francis Group
Boca Raton   London   New York

CRC Press is an imprint of the
Taylor & Francis Group, an **informa** business

A CHAPMAN & HALL BOOK

CRC Press
Taylor & Francis Group
6000 Broken Sound Parkway NW, Suite 300
Boca Raton, FL 33487-2742

First issued in paperback 2017

© 2012 by Taylor & Francis Group, LLC
CRC Press is an imprint of Taylor & Francis Group, an Informa business

No claim to original U.S. Government works

ISBN 13: 978-1-138-11437-1 (pbk)
ISBN 13: 978-1-4398-1273-0 (hbk)

Library of Congress Cataloging-in-Publication Data

Pankratius, Victor.
 Fundamentals of multicore software development / Victor Pankratius, Ali-Reza Adl-Tabatabai, and Walter Tichy.
   p. cm. -- (Chapman & Hall/CRC computational science)
  ISBN 978-1-4398-1273-0 (alk. paper)
  1. Parallel programming (Computer science) 2. Computer software--Development. 3. Multiprocessors. 4. Systems on a chips. I. Adl-Tabatabai, Ali-Reza. II. Tichy, Walter F. III. Title.

QA76.642.P325 2012
005.2'75--dc23   2011041489

Visit the Taylor & Francis Web site at
http://www.taylorandfrancis.com

and the CRC Press Web site at
http://www.crcpress.com

Contents

Foreword

Parallel computing is almost as old as computing itself, but until quite recently it had been interesting only to a small cadre of aficionados. Today, the evolution of technology has elevated it to central importance. Many "true believers" including me were certain it was going to be vital to the mainstream of computing eventually. I thought its advent was imminent in 1980. What stalled its arrival was truly spectacular improvements in processor performance, due partly to ever-faster clocks and partly to instruction-level parallelism. This combination of ideas was abetted by increasingly abundant and inexpensive transistors and was responsible for postponing the need to do anything about the von Neumann bottleneck—the requirement that the operations in a program must appear to execute in linear order—until now.

What some of us discovered about parallel computing during the backwater period of the 1980s formed a foundation for our current understanding. We knew that operations that are independent of each other can be performed in parallel, and so dependence became a key target for compiler analysis. Variables, formerly a benign concept thanks to the von Neumann bottleneck, became a major concern for some of us because of the additional constraints needed to make variables work. Heterodox programming models such as synchronous languages and functional programming were proposed to mitigate antidependences and data races. There was a proliferation of parallel languages and compiler technologies aimed at the known world of applications, and very few computational problems escaped the interest and zeal of those of us who wanted to try to run everything in parallel.

Soon thereafter, though, the worlds of parallel servers and high-performance computing emerged, with architectures derived from interconnected workstation or personal computer processors. The old idea of totally rethinking the programming model for parallel systems took a backseat to pragmatism, and existing languages like Fortran and C were augmented for parallelism and pressed into service. Parallel programming got the reputation of being difficult, even becoming a point of pride for those who did it for a living. Nevertheless, this body of practice became parallel computing in its current form and dominated most people's thinking about the subject until recently.

With the general realization that the von Neumann bottleneck has arrived at last, interest in parallelism has exploded. New insights are emerging as the whole field of computing engages with the challenges. For example, we have an emerging understanding of many common patterns in parallel algorithms and can talk about parallel programming from a new point of view. We understand that a transaction, i.e., an isolated atomic update of the set of variables that comprise the domain of an invariant, is the key abstraction needed to maintain the commutativity of variable updates.

Language innovations in C++, Microsoft .NET, and Java have been introduced to support task-oriented as well as thread-oriented programming. Heterogeneous parallel architectures with both GPUs and CPUs can deliver extremely high performance for parallel programs that are able to exploit them.

Now that most of the field is engaged, progress has been exciting. But there is much left to do. This book paints a great picture of where we are, and gives more than an inkling of where we may go next. As we gain broader, more general experience with parallel computing based on the foundation presented here, we can be sure that we are helping to rewrite the next chapter—probably the most significant one—in the amazing history of computing.

Burton J. Smith
Seattle, Washington

Editors

Dr. Victor Pankratius heads the Multicore Software Engineering investigator group at the Karlsruhe Institute of Technology, Karlsruhe, Germany. He also serves as the elected chairman of the Software Engineering for Parallel Systems (SEPARS) international working group. Dr. Pankratius' current research concentrates on how to make parallel programming easier and covers a range of research topics, including auto-tuning, language design, debugging, and empirical studies. He can be contacted at http://www.victorpankratius.com

Ali-Reza Adl-Tabatabai is the director of the Programming Systems Laboratory and a senior principal engineer at Intel Labs. Ali leads a group of researchers developing new programming technologies and their hardware support for future Intel architectures. His group currently develops new language features for heterogeneous parallelism, compiler and runtime support for parallelism, parallel programming tools, binary translation, and hardware support for all these. Ali holds 37 patents and has published over 40 papers in leading conferences and journals. He received his PhD in computer science from Carnegie Mellon University, Pittsburgh, Pennsylvania, and his BSc in computer science and engineering from the University of California, Los Angeles, California.

Walter F. Tichy received his PhD at Carnegie-Mellon University, Pittsburgh, Pennsylvania, in 1980, with one of the first dissertations on software architecture. His 1979 paper on the subject received the SIGSOFT Most Influential Paper Award in 1992. In 1980, he joined Purdue University, where he developed the revision control system (RCS), a version management system that is the basis of CVS and has been in worldwide use since the early 1980s. After a year at an AI-startup in Pittsburgh, he returned to his native Germany in 1986, where he was appointed chair of programming systems at the University Karlsruhe (now Karlsruhe Institute of Technology).

He is also a director of FZI, a technology transfer institute.

Professor Tichy's research interests include software engineering and parallel computing. He has pioneered a number of new software tools, such as smart recompilation, analysis of software project repositories, graph editors with automatic layout, automatic configuration inference, and language-aware differencing and merging. He has courted controversy by insisting that software researchers need to test their claims with empirical studies rather than rely on intuition or argumentation. He has conducted controlled experiments testing the influence of type-checking, inheritance depth, design patterns, testing methods, and agile methods on programmer productivity.

Professor Tichy has worked with a number of parallel machines, beginning with C.mmp in the 1970s. In the 1990s, he and his students developed Parastation, a communication and management software for computer clusters that made it to the Top500 list of the world's fastest computers several times (rank 10 as of June 2009). Now that multicore chips make parallel computing available to everyone, he is researching tools and methods to simplify the engineering of general-purpose, parallel software. Race detection, auto-tuning, and high-level languages for expressing parallelism are some of his current research efforts.

Contributors

Ali-Reza Adl-Tabatabai
Intel Corporation
Santa Clara, California

David I. August
Princeton University
Princeton, New Jersey

Judith Bishop
Microsoft Research
Redmond, Washington

Hans Boehm
HP Labs
Palo Alto, California

Barbara Chapman
University of Houston
Houston, Texas

Pradeep Dubey
Intel Labs
Santa Clara, California

Michael Garland
NVIDIA Corporation
Santa Clara, California

Vinod Grover
NVIDIA Corporation
Santa Clara, California

Tim Harris
Microsoft Research
Cambridge, United Kingdom

Jialu Huang
Princeton University
Princeton, New Jersey

Thomas B. Jablin
Princeton University
Princeton, New Jersey

Christoph W. Kessler
Linköping University
Linköping, Sweden

Hanjun Kim
Princeton University
Princeton, New Jersey

James LaGrone
University of Houston
Houston, Texas

Thomas R. Mason
Princeton University
Princeton, New Jersey

Tim Mattson
Intel Corporation
DuPont, Washington

Victor Pankratius
Karlsruhe Institute of Technology
Karlsruhe, Germany

Prakash Prabhu
Princeton University
Princeton, New Jersey

Arun Raman
Princeton University
Princeton, New Jersey

Christoph A. Schaefer
Agilent Technologies
Waldbronn, Germany

Kevin Skadron
University of Virginia
Charlottesville, Virginia

Walter F. Tichy
Karlsruhe Institute of Technology
Karlsruhe, Germany

Barry Wilkinson
University of North Carolina, Charlotte
Charlotte, North Carolina

Yun Zhang
Princeton University
Princeton, New Jersey

Chapter 1

Introduction

Victor Pankratius, Ali-Reza Adl-Tabatabai, and Walter F. Tichy

Contents

1.1 Where We Are Today

Multicore chips are about to dramatically change software development. They are already everywhere; in fact, it is difficult to find PCs with a single, main processor. As of this writing, laptops come equipped with two to eight cores. Even smartphones and tablets contain multicore chips. Intel produces chips with 48 cores, Tilera with 100, and Nvidia's graphical processor chips provide several hundred execution units. For major chip manufacturers, multicore has already passed single core in terms of volume shipment. The question for software developers is what to do with this embarrassment of riches.

Ignoring multicore is not an option. One of the reasons is that single processor performance is going to increase only marginally in the future; it might even decrease for lowering energy consumption. Thus, the habit of waiting for the next processor generation to increase application performance no longer works. Future increases of computing power will come from parallelism, and software developers need to embrace parallel programming rather than resist it.

Why did this happen? The current sea change from sequential to parallel processing is driven by the confluence of three events. The first event is the end of exponential growth in single processor performance. This event is caused by our inability to increase clock frequencies without increasing power dissipation. In the past, higher clock speeds could be compensated by lower supply voltages. Since this is no longer

1

possible, increasing clock speeds would exceed the few hundred watts per chip that can practically be dissipated in mass-market computers as well as the power available in battery-operated mobile devices.

The second event is that parallelism internal to the architecture of a processor has reached a point of diminishing returns. Deeper pipelines, instruction-level parallelism, and speculative execution appear to offer no opportunity to significantly improve performance.

The third event is really a continuing trend: Moore's law projecting an exponential growth in the number of transistors per chip continues to hold. The 2009 *International Technology Roadmap for Semiconductors* (http://www.itrs.net/Links/2009ITRS/Home2009.htm) expects this growth to continue for another 10 years; beyond that, fundamental limits of CMOS scaling may slow growth.

The net result is that hardware designers are using the additional transistors to provide additional cores, while keeping clock rates constant. Some of the extra processors may even be specialized, for example, for encryption, video processing, or graphics. Specialized processors are advantageous in that they provide more performance per watt than general-purpose CPUs. Not only will programmers have to deal with parallelism, but also with heterogeneous instruction sets on a single chip.

1.2   How This Book Helps

This book provides an overview of the current programming choices for multicores, written by the leading experts in the field. Since programmers are the ones that will put the power of multicores to work, it is important to understand the various options and choose the best one for the software at hand.

What are the current choices? All mainstream programming languages (C++, Java, .NET, OpenMP) provide threads and synchronization primitives; these are all covered in this book. Parallel programming patterns such as pipelines and thread pools are built upon these primitives; the book discusses Intel's Threading Building Blocks, Microsoft's Task Parallel Library, and Microsoft's PLINQ, a parallel query language.

Graphics processing units (GPUs) require specialized languages; CUDA is the example chosen here. Combined with a chapter on IBM's Cell, the reader can get an understanding of how heterogeneous multicores are programmed.

As to future choices, additional chapters introduce automatic extraction of parallelism, auto-tuning, and transactional memory. The final chapter provides a survey of future applications of multicores, such as recognition, mining, and synthesis.

Most of today's multicore platforms are shared memory systems. A particular topic is conspicuously absent: distributed computing with message passing. Future multicores may well change from shared memory to distributed memory, in which case additional programming techniques will be needed.

1.3 Audience

This book targets students, researchers, and practitioners interested in parallel programming, as well as instructors of courses in parallelism. The authors present the basics of the various parallel programming models in use today, plus an overview of emerging technologies. The emphasis is on software; hardware is only covered to the extent that software developers need to know.

1.4 Organization

- Part I: Basics of Parallel Programming (Chapters 2 and 3)

- Part II: Programming Languages for Multicore (Chapters 4 through 6)

- Part III: Programming Heterogeneous Processors (Chapters 7 and 8)

- Part IV: Emerging Technologies (Chapters 9 through 12)

1.4.1 Part I: Basics of Parallel Programming

In Chapter 2, Barry Wilkinson presents the fundamentals of multicore hardware and parallel programming that every software developer should know. It also talks about fundamental limitations of sequential computing and presents common classifications of parallel computing platforms and processor architectures. Wilkinson introduces basic notions of processes and threads that are relevant for the understanding of higher-level parallel programming in the following chapters. The chapter also explains available forms of parallelism, such as task parallelism, data parallelism, and pipeline parallelism. Wilkinson concludes with a summary of key insights.

In Chapter 3, Tim Mattson presents how the concept of design patterns can be applied to parallel programming. Design patterns provide common solutions to recurring problems and have been used successfully in mainstream object-oriented programming. Applying patterns to parallel programming helps programmers cope with the complexity by reusing strategies that were successful in the past. Mattson introduces a set of design patterns for parallel programming, which he categorizes into software structure patterns and algorithm strategy patterns. The end of the chapter surveys with work in progress in the pattern community.

1.4.2 Part II: Programming Languages for Multicore

In Chapter 4, Hans Boehm shows how C++, one of the most widely used programming languages, supports parallelism. He describes how C++ started off with platform-dependent threading libraries with ill-defined semantics. He discusses the new C++0x standard that carefully specifies the semantics of parallelism in C++ and the rationale for adding threads directly to the language specification. With these extensions, parallel programming in C++ becomes less error prone, and C++ implementations become more robust. Boehm's chapter concludes with a comparison of the current standard to earlier standards.

In Chapter 5, Judy Bishop describes parallelism in .NET and Java. The chapter starts with a presentation of .NET. Bishop outlines particular features of the Task Parallel Library (TPL) and the Parallel Language Integrated Queries (PLINQ), a language that allows declarative queries into datasets that execute in parallel. She also presents examples on how to use parallel loops and futures. Then she discusses parallel programming in Java and constructs of the java.util.concurrent library, including thread pools, task scheduling, and concurrent collections. In an outlook, she sketches proposals for Java on fork-join parallelism and parallel array processing.

In Chapter 6, Barbara Chapman and James LaGrone overview OpenMP. The chapter starts with describing the basic concepts of how OpenMP directives parallelize programs in C, C++, and Fortran. Numerous code examples illustrate loop-level parallelism and task-level parallelism. The authors also explain the principles of how an OpenMP compiler works. The chapter ends with possible future extensions of OpenMP.

1.4.3   Part III: Programming Heterogeneous Processors

In Chapter 7, Michael Garland, Vinod Grover, and Kevin Skandron discuss scalable manycore computing with CUDA. They show how throughput-oriented computing can be implemented with GPUs that are installed in most systems along multicore CPUs. The chapter presents the basics of the GPU machine model and how to program GPUs in the CUDA language extensions for C/C++. In particular, the reader is introduced to parallel compute kernel design, synchronization, task coordination, and memory management handling. The chapter also shows detailed programming examples and gives advice for performance optimization.

In Chapter 8, Christoph Kessler discusses programming approaches for the Cell processor, which is a heterogeneous processor built into Sony's PlayStation® 3. Kessler first outlines the hardware architecture of the Cell processor. Programming approaches for the Cell are introduced based on IBM's software development kit. In particular, Kessler discusses constructs for thread coordination, DMA communication, and SIMD parallelization. The chapter also provides an overview of compilers, libraries, tools, as well as relevant algorithms for scientific computing, sorting, image processing, and signal processing. The chapter concludes with a comparison of the Cell processor and GPUs.

1.4.4   Part IV: Emerging Technologies

In Chapter 9, David I. August, Jialu Huang, Thomas B. Jablin, Hanjun Kim, Thomas R. Mason, Prakash Prabhu, Arun Raman, and Yun Zhang introduce techniques for automatic extraction of parallelism from sequential code. The chapter thoroughly describes dependence analysis as a central building block for automatic parallelization. Numerous examples are used to explain compiler auto-parallelization techniques, such as automatic loop parallelization, speculation, and pipelining techniques. The chapter concludes by discussing the role of auto-parallelization techniques and future developments.

In Chapter 10, Christoph Schaefer, Victor Pankratius, and Walter F. Tichy introduce the basics of automatic performance tuning for parallel applications. The chapter presents a classification of auto-tuning concepts and explains how to design tunable parallel applications. Various techniques are illustrated on how programmers can specify tuning-relevant information for an auto-tuner. The principles of several well-known auto-tuners are compared. An outlook on promising extensions for auto-tuners ends this chapter.

In Chapter 11, Tim Harris presents the basics of the transactional memory programming model, which uses transactions instead of locks. The chapter shows how transactional memory is used in parallel programs and how to implement it in hardware and software. Harris introduces a taxonomy that highlights the differences among existing transactional memory implementations. He also discusses performance issues and optimizations.

In Chapter 12, Pradeep Dubey elaborates on emerging applications that will benefit from multicore systems. He provides a long-term perspective on the most promising directions in the areas of recognition, mining, and synthesis, showing how such applications will be able to take advantage of multicore processors. He also outlines new opportunities for enhanced interactivity in applications and algorithmic opportunities in data-centric applications. Dubey details the implications for multicore software development based on a scalability analysis for various types of applications. The chapter concludes by highlighting the unprecedented opportunities that come with multicore.

Part I

Basics of Parallel Programming

Chapter 2

Fundamentals of Multicore Hardware and Parallel Programming

Barry Wilkinson

Contents

2.1 Introduction

In this chapter, we will describe the background to multicore processors, describe their architectures, and lay the groundwork for the remaining of the book on programming these processors. Multicore processors integrate multiple processor cores on the same integrated circuit chip (die), which are then used collectively to achieve higher overall performance. Constructing a system with multiple processors and using them collectively is a rather obvious idea for performance improvement. In fact, it became evident in the early days of computer design as a potential way of increasing the speed of computer systems. A computer system constructed with multiple processors that are intended to operate together is called a *parallel computer* historically, and programming the processors to operate together is called *parallel programming*. Parallel computers and parallel programming have a long history. The term parallel programming is used by Gill in 1958 (Gill 1958), and his definition of parallel programming is essentially the same as today.

In this chapter, we will first explore the previous work and will start by establishing the limits for performance improvement of processors operating in parallel to satisfy ourselves that there is potential for performance improvement. Then, we will look at the different ways that a system might be constructed with multiple processors. We continue with an outline of the improvements that have occurred in the design of the processors themselves, which have led to enormous increase in the speed of individual processors. These improvements have been so dramatic that the added complexities of parallel computers have limited their use mostly to very high-performance computing in the past. Of course, with improvements in the individual processor, so parallel computers constructed with them also improve proportionately. But programming the multiple processors for collective operation is a challenge, and most demands outside scientific computing have been satisfied with single processor computers, relying on the ever-increasing performance of processors. Unfortunately, further improvements in single processor designs hit major obstacles in the early 2000s, which we will outline. These obstacles led to the multicore approach. We describe architectural designs for a multicore processor and conclude with an outline of the methods for programming multicore systems as an introduction to subsequent chapters on multicore programming.

2.2   Potential for Increased Speed

Since the objective is to use multiple processors collectively for higher performance, before we can explore the different architectures structures one might device, let us first establish the potential for increased speed. The central question is how much faster does the multiprocessor system perform over a single processor system. This can be encapsulated in the speedup factor, $S(p)$, which is defined as

$$S(p) = \frac{\text{Execution time using a single processor (with the best sequential algorithm)}}{\text{Execution time using a multiprocessor system with } p \text{ processors}}$$

$$= \frac{t_s}{t_p} \tag{2.1}$$

where

   t_s is the execution time on a single processor
   t_p is the execution time on system with p processors

Typically, the speedup factor is used to evaluate the performance of a parallel algorithm on a parallel computer. In the comparison, we should use the best-known sequential algorithm using a single processor because the parallel algorithm is likely not to perform as well on a single processor. The execution times could be empirical, that is, measured on real computers. It might be measured by using the Linux time command. Sometimes, one might instrument the code with routines that return wall-clock time, one at the beginning of a section of code to record the start time, and one at the end to record the end time. The elapsed time is the difference. However, this method can be inaccurate as the system usually has other processes executing concurrently in a time-shared fashion. The speedup factor might also be computed theoretically from the number of operations that the algorithms perform. The classical way of evaluating sequential algorithms is by using the time complexity notation, but it is less effective for a parallel algorithm because of uncertainties such as communication times between cooperating parallel processes.

A speedup factor of p with p processors is called *linear speedup*. Conventional wisdom is that the speedup factor should not be greater than p because if a problem is divided into p parts each executed on one processor of a p-processor system and $t_p < t_s/p$, then the same parts could be executed one after the other on a single processor system in time less than t_s. However, there are situations where the speedup factor is greater than p (*superlinear speedup*). The most notable cases are

- When the processors in the multiprocessor system have more memory (cache or main memory) than the single processor system, which provides for increased performance.

- When the multiprocessor has some special feature not present in the single processor system, such as special instructions or hardware accelerators.

- When the algorithm is nondeterministic and happens to provide the solution in one of the parallel parts very quickly, whereas the sequential solution needs to go through many parts to get to the one that has the solution.

The first two cases are not fair hardware comparisons, whereas the last certainly can happen but only with specific problems.

In 1967, Amdahl explored what the maximum speed up would be when a sequential computation is divided into parts and these parts are executed on different processors. This not comparing the best sequential algorithm with a particular parallel algorithm—it is comparing a particular computation mapped onto a single computer and mapped onto a system having multiple processors. Amdahl also assumed that a computation has sections that cannot be divided into parallel parts and these must be performed sequentially on a single processor, and other sections that can be divided equally among the available processors. The sections that cannot be divided into parallel parts would typically be an initialization section of the code and a final section of the code, but there may be several indivisible parts. For the purpose of the analysis, they are lumped together into one section that must be executed sequentially and one section that can be divided into p equal parts and executed in parallel. Let f be the fraction of the whole computation that must be executed sequentially, that is, cannot be divided into parallel parts. Hence, the fraction that can be divided into parts is $1 - f$. If the whole computation executed on a single computer in time t_s, ft_s is indivisible

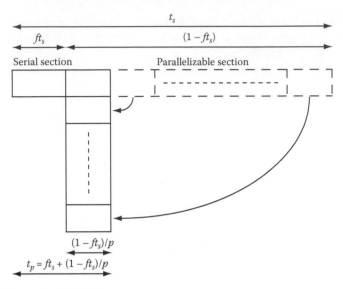

FIGURE 2.1:   Amdahl's law.

and $(1 - f)t_s$ is divisible. The ideal situation is when the divisible section is divided equally among the available processors and then this section would be executed in time $(1 - f)t_s/p$ given p processors. The total execution time using p processors is then $ft_s + (1 - f)t_s/p$ as illustrated in Figure 2.1. The speedup factor is given by

$$S(p) = \frac{t_s}{ft_s + (1 - f)t_s/p} = \frac{1}{f + (1 - f)/p} = \frac{p}{1 + (p - 1)f} \qquad (2.2)$$

This famous equation is known as Amdahl's law (Amdahl 1967). The key observation is that as p increases, $S(p)$ tends to and is limited to $1/f$ as p tends to infinity. For example, suppose the sequential part is 5% of the whole. The maximum speed up is 20 irrespective of the number of processors. This is a very discouraging result. Amdahl used this argument to support the design of ultrahigh speed single processor systems in the 1960s.

Later, Gustafson (1988) described how the conclusion of Amdahl's law might be overcome by considering the effect of increasing the problem size. He argued that when a problem is ported onto a multiprocessor system, larger problem sizes can be considered, that is, the same problem but with a larger number of data values. The starting point for Gustafson's law is the computation on the multiprocessor rather than on the single computer. In Gustafson's analysis, the parallel execution time is kept constant, which we assume to be some acceptable time for waiting for the solution. The computation on the multiprocessor is composed of a fraction that is computed sequentially, say f', and a fraction that contains parallel parts, $(1 - f')$.

This leads to Gustafson's so-called *scaled speedup fraction*, $S'(p)$, given by

$$S'(p) = \frac{f't_p + (1 - f')pt_p}{t_p} = p + (1 - p)f' \qquad (2.3)$$

The fraction, f', is the fraction of the computation on the multiprocessor that cannot be parallelized. This is different to f previously, which is the fraction of the computation on a single computer that cannot be parallelized. The conclusion drawn from Gustafson's law is that it should be possible to get high speedup if we scale up the problem size. For example, if f' is 5%, the scaled speedup computes to 19.05 with 20 processors, whereas with Amdahl's law with $f = 5$%, the speedup computes to 10.26. Gustafson quotes results obtained in practice of very high speedup close to linear on a 1024-processor hypercube.

Others have explored speedup equations over the years, and it has reappeared with the introduction of multicore processors. Hill and Marty (2008) explored Amdahl's law with different architectural arrangements for multicore processors. Woo and Lee (2008) continued this work by considering Amdahl's law and the architectural arrangements in the light of energy efficiency, a key aspect of multicore processors. We will look at the architectural arrangements for multicore processors later.

2.3   Types of Parallel Computing Platforms

If we accept that it should be worthwhile to use a computer with multiple processors, the next question is how should such a system be constructed. Many parallel computing systems have been designed since the 1960s with various architectures. One thing they have in common is they are stored-program computers. Processors execute instructions from a memory and operate upon data. Flynn (1966) created a classification based upon the number of parallel instruction streams and number of data streams:

- Single instruction stream-single data stream (SISD) computer

- Multiple instruction stream-multiple data stream (MIMD) computer

- Single instruction stream-multiple data stream (SIMD) computer

- Multiple instruction stream-single data stream (MISD) computer

A sequential computer has a single instruction stream processing a single data stream (SISD). A general-purpose multiprocessor system comes under the category of a multiple instruction stream-multiple data stream (MIMD) computer. Each processor has its own instruction stream processing its own data stream. There are classes of problems that can be tackled with a more specialized multiprocessor structure able to perform the same operation on multiple data elements simultaneously. Problems include low-level image processing in which all the picture elements (pixels) need to be altered using the same calculation. Simulations of 2D and 3D structures can involve processing all the elements in the solution space using the same calculation. A single instruction stream-multiple data stream (SIMD) computer is designed specifically for executing such applications efficiently. Instructions are provided to perform a specified operation on an array of data elements simultaneously. The operation might be to add a constant to each data element, or multiply data elements. In a SIMD computer, there are multiple processing elements but a single program. This type of design is very efficient for the class of problems it addresses, and some very large SIMD computers have been designed over the years, perhaps the first being the Illiac IV in 1972. The SIMD approach was adopted by supercomputer manufacturers, most notably by Cray computers. SIMD computers are sometimes referred to as vector computers as the SIMD instructions operate upon vectors. SIMD computers still need SISD instructions to be able to construct a program.

SIMD instructions can also be incorporated into regular processors for those times that appropriate problems are presented to it. For example, the Intel Pentium series, starting with the Pentium II in 1996, has SIMD instructions, called MMX (Multi-Media eXtension) instructions, for speeding up multimedia applications. This design used the existing floating-point registers to pack multiple data items (eight bytes, four 16-bit numbers, or two 32-bit numbers) that are then operated upon by the same operation. With the introduction of the Pentium III in 1999, Intel added further SIMD instructions, called SSE (Streaming SIMD extension), operating upon eight new 128-bit registers. Intel continued adding SIMD instructions. SSE2 was first introduced with the Pentium 4, and subsequently, SSE3, SSE4, and SSE5 appeared. In 2008, Intel announced AVX (Advanced Vector extensions) operating upon registers extended to 256 bits. Whereas, large SIMD computers vanished in the 1990s because they could not compete with general purpose multiprocessors (MIMD), SIMD instructions still continue for certain applications. The approach can also be found in graphics cards.

General-purpose multiprocessor systems (MIMD computers) can be divided into two types:

1. Shared memory multiprocessor

2. Distributed memory multicomputer

Shared memory multiprocessor systems are a direct extension of single processor system. In a single processor system, the processor accesses a main memory for program instructions and data. In a shared memory multiprocessor system, multiple processors are arranged to have access to a single main memory. This is a very convenient configuration from a programming prospective as then data generated by one processor and stored in the main memory is immediately accessible by other processors. As in a single processor system, cache memory is present to reduce the need to access main memory continually, and commonly two or three levels of cache memory. However, it can be difficult to scale shared memory systems for a large number of processors because the connection to the common memory becomes a bottleneck. There are several possible programming models for a shared memory system. Mostly, they revolve around using threads, which are independent parallel code sequences within a process. We shall look at the thread programming model in more detail later. Multicore processors, at least with a small number of cores, usually employ shared memory configuration.

Distributed memory is an alternative to shared memory, especially for larger systems. In a distributed memory system, each processor has its own main memory and the processor-memory pair operates as an individual computer. Then, the computers are interconnected. Distributed memory systems have spawned a large number of interconnection networks, especially in the 1970s and 1980s. Notable networks in that era include

- 2D and 3D Mesh networks with computing nodes connected to their nearest neighbor in each direction.

- Hypercube network—a 3D (binary) hypercube is a cube of eight nodes, one at each corner. Each node connects to one other node in each dimension. This construction can be extended to higher dimensions. In an n-dimensional binary hypercube, each node connects to n other nodes, one in each dimension.

- Crossbar switch network—nodes connect to all other nodes, each through a single switch (N^2 switches with N nodes, including switches to themselves).

- Multiple bus network—an extension of a bus architecture in which more than one bus is provided to connect components.

- Tree (switching) network—A network using a tree construction in which the vertices are switches and the nodes are at the leaves of the tree. A path is made through the tree to connect one node to another node.

- Multistage interconnection network—a series of levels of switches make a connection from one node to another node. There are many types of these networks characterized by how the switches are interconnected between levels. Originally, multistage interconnection networks were developed for telephone exchanges. They have since been used in very large computer systems to interconnect processors/computers.

Beginning in the late 1980s, it became feasible to use networked computers as a parallel computing platform. Some early projects used existing networked laboratory computers. In the 1990s, it became cost-effective to interconnect low-cost commodity computers (PCs) with commodity interconnects (Ethernet) to form a high-performance computing cluster, and this approach continues today. The programming model for such a distributed memory system is usually a message-passing model in which messages pass information between the computers. Generally, the programmer inserts message-passing routines in their code. The most widely used suite of message-passing libraries for clusters is MPI. With the advent of multicore computer systems, a cluster of multicore computers can form very high-performance computing platform. Now the programming model for such a cluster may be a hybrid model with threads on each multicore system and message passing between systems. For example, one can use both OpenMP for creating threads and MPI for message passing easily in the same C/C++ program.

Distributed memory computers can also extend to computers that are not physically close. Grid computing refers to a computing platform in which the computers are geographically distributed and interconnected (usually through the Internet) to form a collaborative resource. Grid computing tends to focus on collaborative computing and resource sharing. For more information on Grid computing, see Wilkinson (2010).

2.4   Processor Design

From the early days of computing, there has been an obvious desire to create the fastest possible processor. Amdahl's law suggested that this would be a better approach than using multiple processors. A central architectural design approach widely adopted to achieve increased performance is by using pipelining. Pipelining involves dividing the processing of an instruction into a series of sequential steps and providing a separate unit within the processor for each step. Speedup comes about when multiple instructions are executed in series. Normally, each unit operates for the same time performing its actions for each instruction as they pass through and the pipeline operates in lock-step synchronous fashion. Suppose there are s pipeline units (stages) and n instructions to process in a series. It takes s steps to process the first instruction. The second instruction completes in the next step, the third in the next step and so on. Hence, n instructions are executed in $s + n - 1$ steps, and the speedup compared to a non-pipeline processor is given by

$$s(p) = \frac{sn}{s + n - 1} \qquad (2.4)$$

assuming the non-pipelined processor must complete all s steps of one instruction before starting the next instruction and the steps take the same time. This is a very approximate comparison as the non-pipelined processor probably can be designed to complete the processing of one instruction in less time than s time steps of the pipelined approach. The speedup will tend to s for large n or tend to n for large s. This suggests that there should be a large number of uninterrupted sequential instructions or a long pipeline. Complex processors such as the Pentium IV can have long pipelines, perhaps up to 22 stages, but long pipelines also incur other problems that have to be addressed such as increased number of dependencies between instructions in the pipeline. Uninterrupted sequences of instructions will depend upon the program and is somewhat limited, but still pipelining is central for high-performance processor design. Pipelining is a very cost-effective solution compared to duplicating the whole processor.

The next development for increased single processor design is to make the processor capable of issuing multiple instructions for execution at the same time using multiple parallel execution units. The general term for such a design is a *superscalar processor*. It relies upon *instruction-level parallelism*, that is, being able to find multiple instructions in the instruction stream that can be executed simultaneously. As one can imagine, the processor design is highly complex, and the performance gains will depend upon how many instructions can actually be executed at the same time. There are other architectural improvements, including register renaming to provide dynamically allocated registers from a pool of registers.

Apart from designs that process a single instruction sequence, increased performance can be achieved by processing instructions from different program sequences switching from one sequence to another. Each sequence is a thread, and the technique is known as *multithreading*. The switching between threads might occur after each instruction (fine-grain multithreading) or when a thread is blocked (coarse-grain multithreading). Fine-grain multithreading suggests that each thread sequence will need its own register file. Interleaving instructions increases the distance of related instructions in pipelines and reduces the effects of instruction dependencies. With advent of multiple-issue processors that have multiple execution units, these execution units can be utilized more fully by processing multiple threads. Such multithreaded processor designs are called *simultaneous multithreading (SMT)* because the instructions of different threads are being executed simultaneously using the multiple execution units. Intel calls their version *hyper-threading* and introduced it in versions of the Pentium IV. Intel limited its simultaneous multithreading design to two threads. Performance gains from simultaneous multithreading are somewhat limited, depending upon the application and processor, and are perhaps in the region 10%–30%.

Up to the early 2000s, the approach taken by manufacturers such as Intel was to design a highly complex superscalar processor with techniques for simultaneous operation coupled with using a state-of-the-art fabrication technology to obtain the highest chip density and clock frequency. However, this approach was coming to an end. With clock frequencies reaching almost 4 GHz, technology was not going to provide a continual path upward because of the laws of physics and increasing power consumption that comes with increasing clock frequency and transistor count.

Power consumption of a chip has a static component (leakage currents) and a dynamic component due to switching. Dynamic power consumption is proportional to the clock frequency, the square of the voltage switched, and the capacitive load (Patterson and Hennessy 2009, p. 39). Therefore, each increase in clock frequency will directly increase the power consumption. Voltages have been reduced as a necessary part of decreased feature sizes of the fabrication technology, reducing the power consumption. As the feature size of the chip decreases, the static power becomes more significant and can be 40% of the total power (Asanovic et al. 2006). By the mid-2000s, it had become increasing difficult to limit the power consumption while improving clock frequencies and performance. Patterson calls this the *power wall*.

Wulf and McKee (1995) identified the *memory wall* as caused by the increasing difference between the processor speed and the memory access times. Semiconductor main memory has not kept up with the increasing speed of processors. Some of this can be alleviated by the use of caches and often nowadays multilevel caches, but still it poses a major obstacle. In addition, the *instruction-level parallelism wall* is caused by the increasing difficulty to exploit more parallelism within an instruction sequence. These walls lead to Patterson's "brick wall":

Power wall + Memory wall + Instruction-Level wall = Brick wall

for a sequential processor. Hence, enter the multicore approach for using the ever-increasing number of transistors on a chip. Moore's law originally predicted that the number of transistors on an integrated circuit chip would double approximately every year, later predicting every two years, and sometimes quoted as doubling every 18 months. Now, a prediction is that the number of cores will double every two years or every fabrication technology.

2.5   Multicore Processor Architectures

2.5.1   General

The term multicore processor describes a processor architecture that has multiple independent execution units (cores). (The term *many-core* indicates a large number of cores are present, but we shall simply use the term multicore.) How does a multicore approach get around Patterson's brick wall? With instruction-level parallelism at its limit, we turn to multiple processors to process separate instruction sequences. If we simply duplicated the processors on a chip and all processors operated together, the power would simply increase proportionally and beyond the limits of the chip. Therefore, power consumption must be addressed. The processor cores have to be made more power efficient. One approach used is to reduce the clock frequency, which can result in more than proportional reduction on power consumption. Although reducing the clock frequency will reduce the computational speed, the inclusion of multiple cores provides the potential for increased combined performance if the cores can be used effectively (a big "if"). All multicore designs use this approach. The complexity of each core can be reduced, that is, not use a processor of an extremely aggressive superscalar design. Power can be conserved by switching off parts of the core that are not being used. Temperature sensors can be used to reduce the clock frequency and cut off circuits if the power exceeds limits.

2.5.2   Symmetric Multicore Designs

The most obvious way to design a multicore processor is to replicate identical processor designs on the die (integrated circuit chip) as many times as possible. This is known as a *symmetric multicore* design. Existing processor designs might be used for each core or designs that are based upon existing designs. As we described, the tendency in processor design has been to make processors complex and superscalar for the greatest performance. One could replicate complex superscalar processors on the chip, and companies such as Intel and AMD have followed this approach as their first entry into multicore products, beginning with dual core.

The processor is of course only one part of the overall system design. Memory is required. Main semiconductor memory operates much slower than a processor (the memory wall), partly because of its size and partly because of the dynamic memory design used for high capacity and lower costs. Hence, high-speed cache memory is added near the processor to hold recently used information that can be accessed much more quickly than from the main memory. Cache memory has to operate much faster than the main memory and uses a different circuit design. Speed, capacity, and cost in

Chip

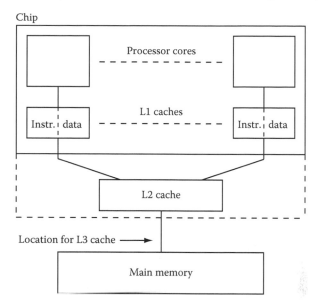

FIGURE 2.2:   Symmetric multicore design.

semiconductor are related. As memory capacity is increased on a memory chip, the tendency is for the device to operate slower for given technology in addition to its cost increasing. This leads to a memory hierarchy with multiple levels of cache memory, that is, an L1 cache, an L2 cache, and possibly an L3 cache, between the processor and main memory. Each level of caches is slower and larger than the previous level and usually includes the information held in the previous level (although not necessarily depending upon the design).

Figure 2.2 shows one possible symmetric multicore design in which each processor core has its own L1 cache fabricated on-chip and close to the core and a single shared external L2 between the multicore chip and the main memory. This configuration would be a simple extension of single processor having an on-chip L1 cache. The L2 cache could also be fabricated on the chip given sufficient chip real estate and an example of such a design is the Intel Core Duo, with two cores on one chip (die) and a shared L2 cache. An L3 cache can be placed between the L2 cache and the main memory as indicated in Figure 2.2 There are several possible variations, including having each core have its own L2 cache and groups of cores sharing an L2 cache on-chip. The Intel Core i7, first released in November 2008, is designed for four, six, or eight cores on the same die. Each core has its own data and instruction L1 caches, its own L2 cache and a shared L3 cache, all on the same die. The cores in the Intel Core i7 also use simultaneously multithreading (two threads per core).

The design shown in Figure 2.2 will not currently scale to a very large number of processors if complex processors are used. However, in a symmetric multicore design, each core need not be a high-performance complex superscalar core. An alternative is to use less complex lower-performance lower-power cores but more of them.

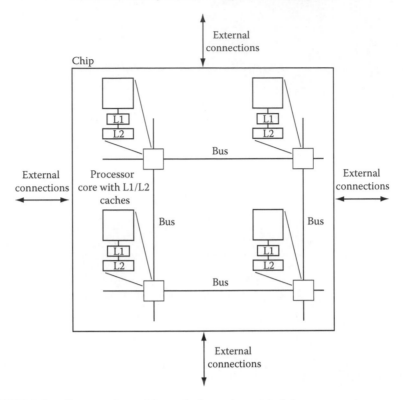

FIGURE 2.3:   Symmetric multicore design using a mesh interconnect.

This approach offers the possibility of fabricating more cores onto the chip, although each core might not be as powerful as a high-performance complex superscalar core. Figure 2.3 shows an arrangement using a 2D bus structure to interconnect the cores. Using a large number of less complex lower-performance lower-power cores is often targeted toward a particular market. An example is the picaChip designed for wireless infrastructure and having 250–300 DSP cores. Another example is TILE64 with 64 cores arranged in a 2D array for networking and digital video processing.

2.5.3   Asymmetric Multicore Designs

In an *asymmetric multicore design*, different cores with different functionality are placed on the chip rather than have one uniform core design. Usually, the configuration is to have one fully functional high-performance superscalar core and large number of smaller less powerful but more power-efficient cores, as illustrated in Figure 2.4. These designs are often targeted toward specific applications. An example is the Sony/Toshiba/IBM Cell Broadband Engine Architecture (cell) used in the PlayStation 3 game console. (Asymmetric design is used in Microsoft's Xbox 360 video game console.) Cell processors are combined with dual-core Opteron

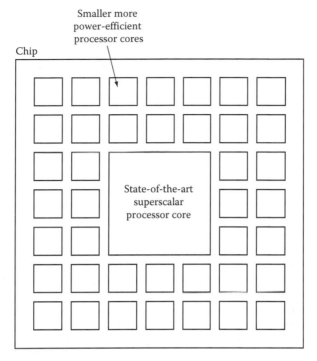

FIGURE 2.4:   Asymmetric multicore design.

processors in a hybrid configuration in the IBM roadrunner supercomputer, which became the world's fastest computer in June 2008 according to the TOP500 ranking.

2.6   Programming Multicore Systems

2.6.1   Processes and Threads

So far, we have just been describing how multicore processors might be designed. Now, we will introduce how such processors might be programmed. A full discussion on multicore programming will be found in subsequent chapters. A multicore processor is generally an implementation of a shared memory multiprocessor system. As mentioned, shared memory multiprocessor systems have been around for many years—just now, the multiple processors are contained within the same die. Hence, the programming techniques used for shared memory multiprocessor systems apply directly to multicore systems. Shared memory systems are programmed to take advantage of the shared memory and to have multiple simultaneously executing program sequences. First, let us review the concept of a *process*, which is used heavily in operating systems to manage multiple executing programs. A process is a complete executable program with its own address space and other dedicated resources (instruction pointer, stack, heap, files, etc.). An operating system will schedule processes for execution, time-sharing the hardware among processes waiting to be executed. This approach enables processes that are stalled because they are waiting for some event such as an I/O transfer to be descheduled and another process be wakened and run on the same processor. On a processor capable of executing only one program sequence at a time, only a single process could be executing at any moment. By switching from one process to another, the system will appear to be executing multiple programs simultaneously although not actually. To differentiate from actual simultaneous operation, we say the processes are executing *concurrently*. On a multiprocessor system, multiple processes could be executing, one on each processor. In that case, we could get true simultaneous executing of the processes. Using the process as the basic unit of simultaneous operation is appropriate for complete but separate programs running on separate computers, for example, in message-passing clusters. The programs might communicate and form a larger parallel program.

A process can be divided into *threads*, separate sequences of code that are intended to be executed concurrently or simultaneously. Being inside the process, these threads share the resources of the process such as memory allocation and files, but each thread needs its own instruction pointer and stack. Creating threads will have much less overhead than creating processes. On a processor capable of executing only one program sequence at a time, threads in a process would be time-shared on the processor, just as processes might be time-shared. However, while processes are completely separate programs doing different tasks, threads within a process are doing tasks associated with the purpose of the process.

The concepts of both processes and threads are embodied in operating systems to enable concurrent operation. Operating system threads are called *kernel threads*. By using such threads, the operating system can switch out threads for other threads when they are stalled. Commonly, threads are assigned a priority number to aid appropriate scheduling, with higher priority threads taking precedence over lower priority threads. Note in the thread model, a process will consist of at least one thread.

The thread programming model has been adopted for programming shared memory multiprocessor systems, including multicore systems. There are two common ways a programmer might create a thread-based program:

1. Using rather low-level thread APIs within a high-level program to explicitly create and manage threads.

2. Using slightly higher-level parallel constructs and directives within a high-level program, which are compiled into thread-based code.

With thread APIs, it will be up to the user to determine exactly what program sequences should be present in each thread and call a "thread create" routine to create the thread from within the main thread or sub-threads. Using higher-level language constructs/directives is easier for the programmer, although it is still up to the programmer to identify what needs to be parallelized into threads. The programmer does not have as much control of the threads and still may need to call some thread information routines. Both ways will be described in detail in subsequent Chapters 4 through 6. Here, we will briefly introduce the basic concepts.

2.6.2   Thread APIs

In the Unix/Linux world, the standard thread API is Pthreads, which was introduced in 1995 in the POSIX (IEEE Portable Operating System Interface) standard under POSIX 1003.1c. Fully compliant POSIX operating systems and most other distributions have Pthreads. A Windows version, called Ptheads-w32, implements most of the Pthread API. In the Windows world, Microsoft also has its own thread API in its C/C++ Windows API (WinAPI) and its .NET framework.

Let us look at facilities provided in threads APIs such as Pthreads and WinAPI. There are differences in the facilities provided by each of these APIs, so we shall discuss general features found in one or the other, or both. Threads are created explicitly by the program by calling a thread-create routine (in both Pthreads and WinAPI). The arguments in the thread-create routine identify the routine that the new thread is to execute and information to be passed to the new thread routine. Threads so created will be destroyed when they reach their end or when an explicit thread exit routine is called (in both Pthreads and WinAPI). The parent thread can be made to wait for the child thread it created to terminate using a thread join routine. When the thread does terminate, it passes status information to the parent thread that created it. It may be that the parent thread does not need to wait for a child thread to terminate. Pthreads offers a version of a thread, called a detached thread, for such a parent–child thread relationship.

Multiple threads can be created using multiple calls to the thread-create routine and joined with multiple calls to the thread join routine. Threads could be created from other threads and threads created on a need basis. It may be that initially it is unknown how many threads are needed. For example, a search algorithm might create new threads as new areas of the search space are discovered. Another example is servicing Web requests. Each request could be allocated a separate thread as each request is received. However, creating new threads incur some overhead. To avoid continually creating and destroying threads, a *thread pool* can be created containing threads waiting to be used. The .NET framework provides explicit support for thread pools, including creating a thread pool and queuing work for threads in the pool.

Threads have access to globally declared shared variables and data, and also each thread will have locally declared variables and data that are only accessibly to that thread for its own use. Programming constructs are necessary to ensure that shared locations cannot be altered simultaneously by multiple threads, which would lead to indeterminate results.

The classical solution to this problem to enshroud the section of the code that is responsible for accessing the location in a so-called *critical section*. Generally, no more than one thread must be in such a critical section at any instant, and this is typically achieved with lock variable that is set to a 1 to indicate a thread is in the critical section and set to a 0 to indicate that no thread is in the critical section. Threads will check the lock is not set before entering the critical section and then set the lock variable if not set and enter the critical section. If the lock is already set (i.e., another thread is in the critical section), the thread has to wait for the critical section to be free. A *spin-lock* simply keeps reading the lock variable in tight loop until the lock variable indicates an unlocked critical section, that is,

```
while (lock == 1) do_nothing;

lock = 1;                        //enter critical section

    critical section

lock = 0;                        // leave critical section
```

A spin lock occupies the thread but is acceptable if it is likely that the critical section will be available soon, otherwise a more efficient solution is needed. When a thread leaves the critical section, it is to reset the lock variable to a 0.

It may be that more than one thread will reach the critical section at the same instant. We must ensure that the actions of each thread in reading and setting the lock variable are not interleaved as this could result in multiple threads setting the lock variable and entering the critical section together. Preventing interleaving is achieved by having instructions that operate in an *atomic* fashion, that is, without interruption by other processors. Most processors are provided with suitable atomic instructions. The Intel Pentium processors can make certain instructions, including the bit test-and-set instruction atomic by prefix the instruction with a LOCK prefix instruction. Apart from using atomic instructions, there are also quite old software algorithms to achieve the same effect such as Dekker's algorithm but mostly nowadays one relies on processor hardware support rather than using software algorithms.

Thread APIs provide support for critical sections. Locks are implemented in so-called *mutually exclusive lock* variables, *mutex*'s, with routines to lock and unlock named mutex's. Another approach is to use *semaphores*. Whereas a lock variable can only be 0 or 1, a semaphore, s, is a positive integer operated upon by two operations, $P(s)$ and $V(s)$. $P(s)$ will wait until s is greater than 0 and then decrement s by 1 and allow the thread to continue. $P(s)$ is used at the beginning of critical section. $V(s)$ will increment s by 1 to release one of the waiting processors. $V(s)$ is used at the end of a critical section. A binary semaphore limits the value of s to 0 or 1 and then behaves in a very similar fashion to a lock, except that semaphores should have an in-build algorithm in $V(s)$ to select wait threads in a fair manner whereas locks may rely upon additional code.

Locks and semaphores are very low-level primitives and can make the programming error-prone. Rather than use explicit lock variables, locks can be associated with objects in an object-oriented language, and this appears in Java. A locking mechanism can be implicit in so-called monitor routines that can only be called by one thread at a time. Java has the synchronized keyword to be used on methods or code sequences to lock them with the associated object lock. In .NET, a section of code can also be protected against more than one thread executing it by the lock keyword.

Threads often need to synchronize between themselves. For example, one thread might need to wait for another thread to create some new data, which it will then consume. A solution is to implement a signaling mechanism in which one thread sends a signal to another thread when an *event* or *condition* occurs. Pthreads uses so-called *condition variables* with a signaling mechanism to indicate that a waiting condition has been achieved. A full treatment of thread APIs in a C/C++ environment is found in Chapter 4. A full treatment of threads in a Java/.NET environment is found in Chapter 5.

2.6.3   OpenMP

OpenMP was developed in the 1990s as a standard for creating shared memory thread-based parallel programs. OpenMP enables the programmer to specify sections of code that are to be executed in parallel. This is done with compiler directives. The compiler is then responsible for creating the individual thread sequences. In addition to a small set of compiler directives, OpenMP has a few supporting routines and environment variables. The number of threads available is set either by a clause in a compiler directive, an explicit routine, or an environment variable. OpenMP is very easy to use but has limitations. Chapter 6 explores OpenMP in detail.

2.7   Parallel Programming Strategies

2.7.1   Task and Data Parallelism

The fundamental tenet of parallel programming is to divide a problem into parts, which are then processed at the same time. There are various strategies one could employ to divide the work into parts. We can first differentiate between the actual computations to be performed and the data that is used in the computations. This leads to two possible decompositions of the problem—either decomposing the computations into parts and executing the parts in parallel (*task parallelism*) or decomposing the data into parts and processing the parts of the data in parallel (*data parallelism*). In data parallelism, the same operation is performed on each part of the data. This form of parallelism has been the basis of machine design as mentioned earlier (SIMD computers). As a programming strategy, data parallelism is particularly relevant if the problem has a large amount of data to process, which is usually the case in scientific computing. It is applicable to problems such as searching a data file or sorting. In practice, decomposition of problems for parallel execution often combines both task parallelism and data parallelism. Decomposition might be done repeatedly and recursively in a divide-and-conquer approach. Subsequent chapters will explore task and data parallelism.

2.7.1.1   Embarrassingly Parallel Computations

The ideal problem to parallelize is one that immediately divides into separate parts, and the parts can be processed simultaneously with no interaction between the parts—they are completely independent on each other. Geoffrey Fox (Wilson 1995) called this type of computation embarrassingly parallel, a term that has found wide acceptance, although the phrase *naturally parallel* is perhaps more apt (Wilkinson and Allen 2005). The implication of embarrassingly parallel computations is that the programmer can immediately see how to divide the work up without interactions between the parts. Usually, there is some interaction at the beginning to start the parts or when initial data is send to the separate parts, and there is usually interaction also at the end to collect results, but if these are the only interactions, we would still describe the problem as embarrassingly parallel. A number of important applications are embarrassingly parallel. In low-level image processing, the picture elements (pixels) of an image can be manipulated simultaneously. Often, all that is needed for each pixel is the initial image or just values of neighboring pixels of the image. Monte Carlo methods can be embarrassingly parallel. They use random selections in numerical calculations. These random selections should be independent of each other, and the calculation based upon them can be done simultaneously. Monte Carlo methods can be used for numerical integration and is especially powerful for problems that cannot be solved easily otherwise. A critical issue for Monte Carlo solutions is the generation of the random selections that can be done in parallel. Traditional random number generators create pseudorandom sequences based upon previously generated numbers and hence are intrinsically sequential. The SPRNG (Scalable Parallel Number Generators) is a library of parallel random number generators that address the issue.

The embarrassingly parallel classification is usually limited to solving a single problem. There are also situations that there is absolutely no interaction, for example, when a complete problem has to be solved repeatedly with different arguments (*parameter sweep*). Such situations are ideal for using multiple processors.

2.7.1.2 Pipelining

We have already mentioned the use of a pipeline in a processor to achieve higher execution speed. The same technique can be used to construct parallel programs. Many problems are constructed as a series of tasks that have to be done in a sequence. This is the basis of normal sequential programming. Multiple processors could be used in a pipeline, one for each task. The output of one task is passed onto the input of the next task in a pipeline. A pipeline can be compared to an assembly line in a factory in which products are assembled by adding parts as they pass down the assembly line. Automobiles are assembled in that way and very efficiently. Multiple automobiles can be assembled at the same time, although only one comes off a single assembly line at a time. Pipelining as a parallel programming strategy is limited to certain applications. A pipeline can be used effectively in parallel processing if the problem can be decomposed into a series of tasks, and there are multiple instances of the problem that need to be solved (c.f. with assembling multiple automobiles). Pipelining can also be used effectively processing a series of data items, each requiring multiple sequential operations to be performed upon them. In that case, the data items are fed down the pipeline.

2.7.1.3 Synchronous Computations

Many problems require synchronization at places during the computation to achieve the desired results. For example, suppose we divide the data set into parts and use the data-parallel approach to operate upon the parts. We generally will need to combine the results from each part before moving to the next stage of the computation. It may be that we will need to perform multiple data-parallel operations. After each one, we have to wait for all operations to complete before moving on. This requires a synchronization mechanism, commonly called a *barrier*. A barrier is inserted into the code at the point that synchronization is to take place and causes all processes to wait for each other before moving on from that point. Barrier mechanisms exist in both message-passing libraries and thread environments. OpenMP has an implicit barrier in its constructs that spawn threads such that all threads must complete before continuing with subsequent code (unless the programmer inserts a nowait clause where allowed).

Synchronous iteration (also called *synchronous parallelism*) describes a program that performs a loop with a body having multiple parallel parts. Each iteration must wait for all the parallel parts in the loop body to complete before embarking on the next iteration. A common example of synchronous iteration is solving a problem by converging on the solution through iteration. A system of linear equations can be solved this way if certain mathematical convergence conditions are satisfied. Similarly, differential equations can be solved by a finite difference method through iteration.

Synchronization in parallel programs will cause significant delay in the computations at each synchronization point (although it certainly makes it easier to debug). For maximum computational speed, every effort should be made to reduce the need for synchronization. In *partially synchronous* or *asynchronous* iterative methods, synchronization is not performed on every iteration (Baudet 1978). Villalobos and Wilkinson (2008) explored this approach with data buffers holding several previously computed values.

2.7.1.4 Workpool

A *workpool* describes a collection of tasks to be handed out to computing resources for execution on a demand basis. It is a very effective way of balancing the load. It can take into account the speed of the compute resources to complete tasks and also when the number of tasks varies. The most basic way to construct a workpool is to start with the tasks that need to be performed in a workpool task queue. Compute resources are then given tasks to perform from this queue. When a compute resource finishes its tasks, it requests further tasks. It may be that processing a task will generate new tasks. These tasks might be returned by the compute resources and placed in the workpool task queue for redistribution. Workpools can be centralized or distributed. A single workpool might hold all the pending tasks, or there could be multiple workpools at different sites to distribute the communication. A fully distributed workpool would have a tasks queue in each compute resource. In the *received-initiated* approach, other compute resources request tasks from compute resources, typically when they have nothing to do. In the *sender-initiated* approach, compute resources choose to pass on tasks to other resources, typically when they are overloaded with work. Counter-intuitively, the receiver-initiated approach has been shown to work well under high overall system load while the sender-initiated approach has been shown to work well under light overall system load. Both methods can be used.

Determining when an overall computation has finished may require some care. With physically distributed compute resources, simply having empty task queues and all tasks completed are insufficient conditions as tasks could be in transition from one resource to another and not yet received. Tasks in transit can be handled by employing a messaging protocol in which an acknowledgment message is sent back to the source for every request message. Only when all acknowledgements have been received can the status of the computation can be established. There are a number of ingenious algorithms for detecting termination of a distributed algorithm, see Wilkinson and Allen (2005).

2.8 Summary

The move to multicore processors has come after a very long history of computer design. The primary motive is to increase the speed of the system. Increasing in the speed of a single processor hit the famous Patterson "brick" wall, which described the combination of not being able to cope with increase of power dissipation, limited prospect for any further instruction-level parallelism, and limits in main memory speeds. Hence, manufacturers have moved from developing a single high-performance processor on one die to having multiple processor cores on one die. This development began with two cores on one die and continues with more cores on one die. The cores may individually be less powerful that might be possible if the whole die was dedicated to a single high-performance superscalar processor, but collectively, they offer significant more computational resources. The problem now is to use these computational resources effectively. For that, we draw upon the work of the parallel programming community.

References

Amdahl, G. 1967. Validity of the single-processor approach to achieving large-scale computing capabilities. *Proc 1967 AFIPS*, vol. 30, New York, p. 483.

Asanovic, K., R. Bodik, B. C. Catanzaro, J. J. Gebis, P. Husbands, K. Keutzer, D. A. Patterson, W. L. Plishker, J. Shalf, S. W. Williams, and K. A. Yelick. 2006. The landscape of parallel computing research: A view from Berkeley. University of California at Berkeley, technical report no. UCB/EECS-2006-183, http://www.eecs.berkeley.edu/Pubs/TechRpts/2006/EECS-2006-183.html

Flynn, M. J. 1966. Very high speed computing systems. *Proceedings of the IEEE* 12:1901–1909.

Gill, S. 1958. Parallel programming. *The Computer Journal* 1(1): 2–10.

Gustafson, J. L. 1988. Reevaluating Amdahl's law. *Communication of the ACM* 31(1): 532–533.

Hill, M. D. and M. R. Marty. 2008. Amdahl's law in the multicore era. *IEEE Computer* 41(7):33–38.

Patterson, D. A. and J. L. Hennessy. 2009. *Computer Organization and Design: The Hardware/Software Interface*, 4th edn. Burlington, MA: Morgan Kaufmann.

Villalobos, J. F. and B. Wilkinson. 2008. Latency hiding by redundant processing: A technique for grid-enabled, iterative, synchronous parallel programs. *15th Mardi Gras Conference*, January 30, Baton Rouge, LA.

Wilkinson. B. 2010. *Grid Computing: Techniques and Applications*. Boca Raton, FL: Chapman & Hall/CRC Computational Science Series.

Wilkinson, B. and M. Allen. 2005. *Parallel Programming: Techniques and Applications Using Networked Workstations and Parallel Computers*, 2nd edn. Upper Saddle River, NJ: Prentice Hall.

Wilson, G. V. 1995. *Practical Parallel Programming*. Cambridge, MA: MIT Press.

Woo, D. H. and H.-H. S. Lee. 2008. Extending Amdahl's law for energy-efficient computing in many-core era. *IEEE Computer* 41(12):24–31.

Wulf, W. and S. McKee. 1995. Hitting the memory wall: Implications of the obvious. *ACM SIGArch Computer Architecture News* 23(1):20–24.

Chapter 3

Parallel Design Patterns

Tim Mattson

Contents

3.1 Parallel Programming Challenge

With multi-core processors in virtually every segment of the computing market, mainstream programmers need to routinely write parallel software. While progress has been made with automatic parallelization (see Chapter 9), experience with parallel computing has shown that in most cases, programmers will need to create parallel software "by hand." This transition from serial to parallel software is a huge challenge for the software industry [Hwu08]; perhaps the biggest it has ever faced.

While parallel computing is new to mainstream programmers, it has a long history in scientific computing. In the 1980s, the scientific computing community was faced with a transition from vector supercomputers supported by vectorizing compilers to explicitly parallel massively parallel processors and clusters. Computer scientists responded by creating countless parallel programming languages, but these for the most part failed. Only a tiny fraction of programmers in the sciences took up parallel programming; and those that did tended to use low-level message passing libraries [MPI] that emphasized portability and control over elegance.

New high-level languages were not enough to solve the parallel programming problem in the 1980s and 1990s, and there is no reason to believe it will be different this time as we help general purpose programmers adopt parallel programming.

It is my hypothesis that the key to solving this problem is to focus on how experienced programmers think about parallel programming. Experienced parallel programmers, once they understand how the concurrency in their problem supports parallel execution, can use any sufficiently complete parallel programming language to create parallel software. Therefore, to solve the parallel programming problem for mainstream programmers, we need to capture how expert parallel programmers think and provide that understanding to mainstream programmers. In other words, we do not need new languages or sophisticated tools (though they help); we need to help mainstream programmers learn to "think parallel."

Learning to reason about parallel algorithms is a daunting task. Fortunately, we have a few decades of experience to draw on. All we need to do is distill this collective wisdom into a form that we can put down in writing and communicate to people new to parallel computing. We believe design patterns are an effective tool to help us accomplish this task.

3.2   Design Patterns: Background and History

What makes a design excellent? Often, the people using a design cannot say what makes it excellent. These users may be able to distinguish between excellence and the mundane, but in most cases, they will not be able to explain what makes an excellent design special. Experts, however, become "experts" by virtue of their ability to recognize excellence in a design, and this helps them to reproduce excellence in their own designs. But how to you capture that essential knowledge to provide the specialized language experts use to communicate among themselves and to help turn novices into experts?

Christopher Alexander [Alexander77] struggled with this problem and hit upon what has proven to be a useful technique. The idea is to study excellent designs and identify the established solutions to recurring problems. Once found, these solutions are given a descriptive name and written down using a carefully structured format. These solutions are called "design patterns." When you consider a complicated problem with many steps, patterns emerge for different facets of the problem. One pattern flows into another creating a web of design patterns to help guide an emerging design. We call this web of interlocking patterns a "design pattern language."

Patterns have a long history in software. If we go back to the early 1990s when object-oriented programming was in its infancy, the field was quite confused. Experts were emerging and quality designs appeared, but there was a great deal of chaos as programmers struggled with how to make object-oriented abstractions work for them. Gamma, Helm, Johnson, and Vlissides (affectionately known as "the gang of four") made a major contribution to object-oriented programming with their famous book on design patterns for object-oriented design [Gamma94]. This book had a transformative effect on object-oriented programming and helped it rapidly transition from chaos into order. It helped novices master object-oriented design and created consistent jargon experts could use to advance the field.

We believe design patterns can have a similar transformative effect on multi-core programming. In the next few sections, we summarize some of the important design patterns in parallel programming. This builds off an earlier pattern language for parallel programming [Mattson04] and more recent work with the ParLab at UC Berkeley [Keutzer10].

3.3 Essential Patterns for Parallel Programming

Patterns are mined from good designs. Much of the history of parallel programming comes from the sciences, so it should be no surprise that we tend to mine patterns from technical applications. As you look closely at the compute intensive applications that will be important on future multi-core processors [Parsec08], however, the basic parallel algorithms are essentially the same as the parallel algorithms form scientific computing. We therefore believe patterns with their roots in the sciences will continue to be important for mainstream programmers working with multi-core chips.

In the following sections, we will summarize some of the key patterns used in designing parallel software. The details are available elsewhere [Mattson04,OPL]. For our purposes, a high level outline of the pattern should suffice. We will consider two classes of design patterns: parallel algorithm strategy patterns and implementation strategy patterns. The *parallel algorithm strategy* patterns address how the concurrency in an algorithm can be organized and how dependencies can be resolved. They lead to a high level plan for a parallel algorithm.

- Task parallelism

- Data parallelism

- Divide and conquer

- Pipeline

- Geometric decomposition

The lower-level *implementation strategy* patterns describe how that algorithm strategy can be supported in software.

- SPMD (single program multiple data)

- SIMD (single instruction multiple data)

- Loop-level parallelism

- Fork-join

- Master-worker

The complete text for these patterns goes well beyond the scope of this survey. Instead, we use an abbreviated pattern-format that exposes the most essential elements of the patterns. For each pattern, we provide the following:

- A descriptive phrase or simple name that succinctly defines the solution addressed by the pattern

- A brief description of the pattern

- A discussion of when to use the pattern

- Detailed information about how to apply the pattern in a software design

- A description of where the pattern is used in practice

- A discussion of related patterns

3.3.1   Parallel Algorithm Strategy Patterns

Parallel programming is the application of concurrency to (1) solve a problem in less time or (2) solve a bigger problem in a fixed amount of time. Concurrency is present in a system when multiple tasks can be active and potentially make progress at the same time. The processes of creating a parallel algorithm can therefore be summarized in two steps:

1. Find tasks that can execute concurrently.

2. Manage data conflicts between tasks so the same results are produced (within round-off error) for every semantically allowed way the execution of the tasks can be interleaved.

To solve any problem in parallel, you must decompose the problem into concurrent tasks AND decompose the data to manage data access conflicts between tasks.

The goal of the algorithm strategy patterns is to define this dual task/data decomposition. They start with a conceptual understanding of a problem and how it can be executed in parallel. They finish with a decomposition of the problem into concurrent tasks and data.

3.3.1.1   Task Parallelism Pattern

The concurrency is expressed directly in terms of an enumerated set of tasks that are either naturally independent or can be transformed into independent tasks.

Use When   The concurrency in a problem may present itself directly in terms of a collection of distinct tasks. The definition of the "tasks" is diverse. They can be a set of files you need to input and process. They can be the rays of light between a camera and a light source in a ray tracing problem. Or they can be a set of operations to be applied to a distinct partition of a graph. In each case, the tasks addressed by this pattern do not depend on each other. Often the independent tasks are a natural consequence of the problem definition. In other cases where this pattern can be used, a transformation of the data or a post-processing step to address dependencies can create the independent tasks needed by this pattern.

Details   Give a set of N tasks

$$t_i \in T(N)$$

the task parallel pattern schedules each task, t_i, for execution by a collection of processing elements. There are three key steps in applying the task parallelism pattern to a problem:

1. The problem must be decomposed into tasks, t_i.

2. Data associated with each task must be decomposed and isolated to the tasks so they can execute independently and in any order.

3. The tasks must be scheduled for execution by a collection of processing elements. The schedule must balance the load among the processing elements so they finish at approximately the same time.

For an effective application of this pattern, there are two criteria the tasks must meet. First, there must be more tasks (often many more) than processing elements in the target platform. Second, the individual tasks must be sufficiently compute-intensive in order to offset the overheads in creating and managing them.

The dependencies between tasks play a key role in how this pattern is applied. In general, the tasks need to be mostly independent from each other. When tasks depend on each other in complicated ways, it is often better to use an approach that expresses the concurrency directly in terms of the data.

There are two common cases encountered when working with the dependencies in a task parallel problem. In the first case, there are no dependencies. In this case, the tasks can execute completely independently from each other. This is often called an "embarrassingly parallel" algorithm. In the second case, the tasks can be transformed to separate the management of the dependencies from the execution of the tasks. Once all the tasks have completed, the dependencies are handled as a distinct phase of the computation. These are called *separable dependencies*.

A key technique that arises in separable-dependency algorithms is "data replication and reduction." Data central to computation over a set of tasks is replicated with one copy per computational agent working on the tasks. Each agent proceeds independently, accumulating results into its own copy of the data. When all the tasks are complete, the agents cooperate to combine (or reduce) their own copies of the data to produce the final result.

Known Uses   This pattern, particularly for the embarrassingly parallel case, may be the most commonly used pattern in the history of parallel computing. Rendering frames in a film, ray tracing, drug docking studies are just a few of the countless instances of this pattern in action. Map-reduce [Dean04] can be viewed as a "separable-dependencies" variant of the task parallel pattern. The computations of short-range forces in molecular dynamics problems often make use of the separable-dependencies version of the task parallel pattern.

Related Patterns   This pattern is related to the divide-and-conquer pattern in that they both address concurrency directly in terms of a set of tasks. In the task parallel pattern, however, the tasks are enumerated directly rather than recursively.

Software structure patterns to support the task parallel pattern are chosen based on the target platform. For shared address space platforms, the loop-level parallelism pattern is an effective choice. The SPMD pattern is straightforward to apply to the task parallel pattern by mapping the set of tasks onto the ID of each processing element. Perhaps the most famous software structure pattern to support the task parallel pattern is the master-worker pattern. This pattern (as we will discuss later) is very effective at balancing the load in a task parallel computation.

3.3.1.2   Data Parallelism

The concurrency is expressed as a single function applied to each member of a set of data elements. In other words, the concurrency is "in the data."

Use When   Problems that utilize this pattern are defined in terms of data with a regular structure that can be exploited in designing a concurrent algorithm. Graphics algorithms are perhaps the most common example since in many cases, a similar (if not identical) operation is applied to each pixel in an image. Physics simulations based on structured grids, a common class of problems in scientific computing but also in computer games, are often addressed with a data parallel pattern. The shared feature of these problems is that each member of a data structure can be updated with essentially the same instructions with few (if any) dependencies between them. The dependencies are of course the key challenge since they too must be managed in essentially the same way for each data element. A good example of a data parallel pattern with dependencies is the relaxation algorithm used in partial differential equation solvers for periodic problem domains. In these problems, a stencil is used to define how each point in the domain is updated. This defines a dependency between points in the domain, but it is the same for each point and hence (other than at the boundaries) can be handled naturally with a data parallel algorithm.

Details   This pattern is trivial in the case of simple vector operations. For example, to add two equal length vectors together in parallel, we can apply a single function (the addition operator) to each pair of vector elements and place the result in the corresponding element of a third vector. For example, using the vector type in OpenCL, we can define four element vectors and add them with a single expression:

```
float4 a = (float4) (2.0, 3.0, 4.0, 5.0);
float4 b = (float4) (3.0, 4.0, 5.0, 2.0);
float4 c;

c = a + b;    // c = (float4) (5.0, 7.0, 9.0, 7.0);
```

The advantage of vector operations is they map directly onto the vector units found in many microprocessors.

The challenge is how do we generalize this trivial case to more complex situations? We do this by defining an abstract index space. The data in the problem is aligned to this index space. Then the concurrency is expressed in terms of the data by running a single stream of operations for each point in the index space.

For example, in an image processing problem applied to an N by M image, we define a rectangular "index space" as a grid of dimension N by M. We map our image, A, onto this index space as well as any other data structures involved in the computation (in this case, a filter array). We then express computations on the image onto points in the index space and define the algorithm as

```
For each point (i,j) in the N by M index space {
    FFT(A, FORWARD)(i,j);
    Construct_filter(B(i,j));
    C(i,j) = A(i,j) * B(i,j);
    FFT(A, INVERSE)(i,j);
}
```

Note that the FFT operations involve complex data movement and operations over collections of points. But the data parallelism pattern is still honored since these complex operations are the same on each subset of data elements. As suggested by this example, the data parallelism pattern can be applied to a wide range of problems, not just those with simple element-wise updates over aligned data.

Known Uses   This pattern is heavily used in graphics, so much so that GPU hardware is specifically designed to support problems that use the data parallelism pattern. Image processing, partial differential equation solvers, and linear algebra problems also map well onto this pattern. More complex applications of this pattern can be found in Chapter 7, Scalable Manycore Computing with CUDA.

Related Patterns   The geometric decomposition pattern is often described as a form of data parallelism. It is not quite the same as this pattern in that the functions applied to each tile in a geometric decomposition pattern usually differ between tiles (i.e., it is not strictly a single function applied to each of the tiles). But it shares the general idea of defining parallelism in terms of concurrent updates to different elements of a larger data structure.

The data parallelism pattern can be implemented in software with the SIMD, loop parallelism, or even SPMD patterns, the choice often being driven by the target platform.

A data parallel pattern can often be transformed into a task parallel pattern. If the single function, t, applied to the members of a set of data elements, d_i, then we can define

$$t_i = t(d_i)$$

and we can address the problem using the task parallelism pattern. Note, however, that the converse relationship does not hold, that is, you cannot generally transform a task parallel pattern into a data parallel pattern. It is for this reason that the task parallelism patterns are considered to be the more general patterns for parallel programming.

3.3.1.3   Divide and Conquer

Concurrency is generated by recursively dividing a problem into smaller and smaller tasks until the tasks are small enough to solve directly.

Use When   This pattern describes a range of closely related algorithms. The common feature of this class of problems is some way to recursively split a problem into smaller subproblems; either by traversing a recursive data structure or through recursion relations that define ways to split a problem. For example, an optimization problem often produces a directed acyclic graph of choices as parameters in a solution space are explored.

Details   The structure of the problem is used to define an operation to split the problem into smaller subproblems. The existence of this structure is fundamental to the solution. These subproblems are split again to produce even smaller problems. This continues recursively until a problem is small enough to solve directly. The recursive splitting can be thought of as splitting tasks into subtasks (e.g., dynamic programming) or in terms of recursive decomposition of data (e.g., the octree decomposition in a Barnes–Hut algorithm). The core ideas behind the solution are basically the same in either case. There are many variations to the divide-and-conquer pattern.

- The classic *divide-and-conquer* algorithm recursively generates the subproblems, solves them, and then explicitly combines them (reversing the recursive splitting process) to construct the global solution.

- A *dynamic programming* algorithm uses a top-down algorithm to generate tasks as part of the recursive splitting but checks tasks to see if they have already been solved (a process called memoization). If there is significant overlap between subproblems, this can result in a significant reduction in total work.

- The *backtrack branch and bound* algorithm is used for search problems and treats the subproblems as branches in a decision tree. For these problems, there is some way to define a bound over the range of the solutions. If a branch is shown to fall outside the current bound, the entire branch can be discarded and the solution can backtrack to consider a new branch.

These are just a few of the possible options with this large class of solutions. An example of the divide-and-conquer pattern is provided in Figure 3.1, where we reduce

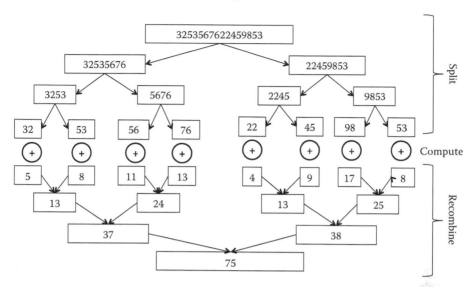

FIGURE 3.1: Divide-and-conquer algorithm for reducing a string of numbers into a single scalar using summation. Three distinct phases are clear: split problem into smaller subproblems, operate at the leaves of the resulting tree, and reverse the "split process" to recombine subproblem solutions into a solution to the global problem.

a string of integers into a single value using summation. Subproblems are generated at each level by splitting the list into two smaller lists. The subproblem is small enough to solve directly once there are a pair of integers at which point the "operate phase" occurs. Finally, the subproblems are recombined (using summations) to generate the global solution.

Known Uses This pattern is heavily used in optimization and decision support problems. Programmers using Cilk [Blumofc95] and the explicit task construct in OpenMP 3.0 (see Chapter 6) make heavy use of this pattern as well.

Related Patterns Implementations of this pattern in software often use the fork-join pattern. To avoid the high costs of explicitly spawning and destroying threads, implementations of this pattern build schedulers to manage tasks forked during the recursion and execute them using a thread pool of some sort.

3.3.1.4 Pipeline

Concurrency produced by passing multiple data elements through a sequence of computational stages.

Use When Concurrency can be difficult to extract from a problem. Sometimes, a problem is composed of a sequence of tasks that must execute in a fixed order. It appears in such cases that there just is not a way to make the program run in parallel. When this occurs, however, it is sometimes the case that there are multiple data elements that must pass through the sequence of tasks. While each of the tasks must complete in order, it may be possible to run each data element as an independent flow through the collection of tasks. This only works if the tasks do not hold any state, i.e., if the tasks take input data, computes a result based on the data, and then passes the result to the next task.

Details   Pipeline problems establish a sequence of stages as stateless filters. Data elements, d_i, pass through the pipeline and a computation is complete after a data element has passed through all of the stages. For example, consider a three stage pipeline with data elements d_i passing through the stages:

```
time = 0   X1(dᵢ)       → X2()        → X3()
time = 1   X1(dᵢ₋₁)     → X2(dᵢ)      → X3()
time = 2   X1(dᵢ₋₂)     → X2(dᵢ₋₁)    → X3(dᵢ)
```

When the computation starts, only the first stage, X1, is active and there is no concurrency. As X1 completes, it passes its result to X2 which then starts its own computation while X1 picks up the next data element. This continues until all the pipeline states are busy.

The amount of concurrency available is equal to the number of stages, and this is only available once the pipeline has been filled. The work prior to filling the pipeline and the diminishing work as the pipeline drains constrains the amount of concurrency these problems can exploit.

This pattern, however, is extremely important. Given that each data element must pass through each stage in order, the options for exploiting concurrency is severely limited in these problems (unless each of the states can be parallelized internally).

Known Uses   This pattern is commonly used in digital signal processing and image processing applications where each unit of a larger multicomponents data set is passed through a sequence of digital filters.

Related Patterns   This pattern is also known as the pipe-and-filter pattern [Shaw95]. This emphasizes that the stages are ideally stateless filters.

The pipeline pattern can be supported by lower-level patterns in a number of ways. The key is to create separate units of executions (i.e., threads or processes) and manage data movement between them. This can be set up with the SPMD or the fork-join patterns. The challenge is to safely manage movement of data between stages and to be prepared for cases where there is an imbalance between stages. In this case, the channels between stages need to be buffered (perhaps using a shared queue data structure).

3.3.1.5   Geometric Decomposition

A variation of data parallelism where a larger data structure is decomposed into tiles which can be updated in parallel. Note, however, that the processing occurring within each tile may be different.

Use When   This pattern is used when the data associated with a program plays a dominant role in how we understand the problem. For example, in an image processing algorithm where a filter is applied to neighborhoods of pixels in the image, the problem is best defined in terms of the image itself. In cases such as these, it is often best to define the concurrency in terms of the data and how it is decomposed into relatively independent blocks.

Details   Define the concurrent tasks in terms of the data decomposition. The data breaks down into blocks. Tasks update blocks of data. When data blocks share boundaries or otherwise depend on each other, the algorithm must take that into account; often adding temporary storage around the boundaries (so called "ghost cells") to hold nonlocal data. In other words, the algorithm is often broken down into three segments:

1. Communicate boundary regions

2. Update interior points in each block

3. Update boundary regions of each block

The split into two update-phases, one for interior points and the other for boundary regions, is done so you can overlap communication and computation steps.

Known Uses   This is one of the most commonly used patterns in scientific computing. Most partial differential equation solvers and linear algebra programs are parallelized using this pattern. For example, in a stencil computation, each point in an array is replaced with a value computed from elements in its neighborhood. If the domain is decomposed into tiles, communication is only required at the boundaries of each tile. All the updates on the interior can be done concurrently. If communication of data on the boundaries of tiles can be carried out during computations in the interior, very high levels of concurrency can be realized using this pattern.

Related Patterns   This pattern is closely related to the data parallelism pattern. In fact, for those cases where each tile is processed by the same task, this problem transforms into an instance of the data parallelism pattern.

This pattern is commonly implemented with the SPMD and loop-level parallelism patterns. In many cases, it can be implemented with the SIMD pattern, but inefficiencies arise since boundary conditions may require masking out subsets of processing elements for portions of the computation (for example, at the boundaries of a domain).

3.3.2   Implementation Strategy Patterns

These patterns describe how threads or processes execute in a program, i.e., they are intimately connected with how an algorithm design is implemented in source code. The input to this layer of the design pattern language is a problem decomposed into concurrent tasks, a data decomposition, and a strategy for how to execute the tasks in parallel. The result of this layer of the pattern language is source code (or pseudocode) implementing an algorithm in software.

3.3.2.1   SPMD

A single program is written that is executed by multiple threads or processes. They use the ID of each thread or process to select different pathways through the code or choose which elements of data structures to operate upon.

Use When   Concurrent tasks run the same code but with different subsets of key data structures or occasional branches to execute different code on different processes or threads.

You intend to use a message passing programming model (i.e., MPI) and treat the system as a distributed memory computer.

Details   The SPMD pattern is possibly the most commonly used pattern in the history of parallel supercomputing. The code running on each process or thread is the same, built from a single program. Variation between tasks is handled through expressions based on the thread or process ID. The variety of SPMD programs is too great to summarize here. We will focus on just a few cases.

In the first case, loop iterations are assigned to different tasks based on the process ID. Assume for this (and following) examples that each thread or process is labeled by a rank ranging from zero to the number of threads or processes minus one.

```
ID = 0, 1, 2, ... (Num_procs − 1)
```

We can use the techniques described in the "loop parallelism pattern" to create independent loop iterations and then change the loop to split up iterations between processes or threads.

```
for(i=ID; i<Num_iterations; i=i+Num_procs)  {...}
```

This approach divides the work between the threads or processes executing the program, but it may lead to programs that do not effectively reuse data within the caches. A better approach is to define blocks of contiguous iterations. For example, working with the geometric decomposition pattern, we can break up the columns of a matrix into distinct sets based on the ID and `num_procs`. For example, given a square matrix of order `Norder`, we can break up columns or matrix into sets one of which is assigned to each process or thread.

```
IStart = (ID* Norder/num_procs)
ILast = (ID + 1) *Norder/num_procs
If  ( ID == (num_procs − 1))  ILast = Norder
for(i=IStart; i<ILast; i++)  { ... }
```

Programming models designed for general purpose programming of GPUs (see Chapter 7, GPU programming or [39]) provide another common instance of the SPMD programming model. Central to these programming models is a data parallel mode of operation. The software framework defines an index space. A kernel is launched for each point in the index space. The kernel follows the common SPMD model; it queries the system to discover its ID and uses this ID to (1) select which point in the index space it will handle and (2) choose a path through the code. For example, the following OpenCL kernel would be used to carry out a vector sum:

```
_kernel void vec_add (_global const float *a,
                      _global const float *b,
                      _global float *c)
{
    int gid = get_global_id(0);
    c[gid] = a[gid] + b[gid];
}
```

A keyword identifies this function as a kernel. The arguments are listed as "_global" to indicate to the compiler that they will be imported from the host onto the compute device. The SPMD pattern can be seen in the body of the function. The kernel queries its ID "gid" and then uses it to select which elements of the array to sum. Notice that in this case, each kernel will execute a single stream of instructions, so in this form, the pattern is identical to the SIMD pattern. We consider this an SPMD pattern, however, since a kernel can branch based on the ID, causing each instance of a kernel's execution to execute a significantly different set of instructions from the body of the kernel function.

Known Uses   This pattern is heavily used by MPI programmers and OpenMP programmers wanting more control over how data is managed between different parts of the program. It is becoming increasingly important as a key pattern in GPGPU programming.

Related Patterns   The SPMD pattern is very general and is used to support software designs based on the data parallel, geometric decomposition, and task parallel patterns.

As mentioned earlier, this pattern is very similar to the SIMD pattern. In fact, the SIMD pattern can be considered a simple case of the more general SPMD pattern.

3.3.2.2   SIMD

One stream of instructions is executed. The system (usually with hardware support) applies this stream of instructions to multiple data elements, thereby supporting concurrent execution.

Use When   Use this pattern for problems that are strictly data parallel, i.e., where the same instruction will be applied to each of a set of data elements. If the application involves a significant degree of branching, this pattern is difficult to apply. This limits the applicability of the pattern, but the benefits are considerable since the SIMD pattern makes it so much easier to reason about concurrency. In a multithreaded program, for example, the programmer needs to think about the program and understand every semantically allowed way the instructions can be interleaved. Validating that every possible way instructions can be interleaved is correct can be prohibitively difficult. With the SIMD pattern, however, there is only one stream of instructions.

Details   This pattern is tightly coupled to the features provided by the platform. For example, the SIMD pattern is used when writing code for the vector units of a microprocessor. Writing code using a native vector instruction set sacrifices portability.

Fortunately, high-level languages are emerging that support portable programming of vector units. For example, consider the vector addition example we discussed earlier. OpenCL includes a number of vector types such as the float4 type.

```
float4 A, B, C;
A = B + C;
```

The single SIMD instruction "A = B + C" applies a pair-wise addition operation to the four elements of B and C writing the results into the analogous positions in A. This approach can be used throughout a program leading to a complex application where a single stream of instructions operating on vector data yields significant amounts of concurrency.

The SIMD pattern would have a limited impact if it was restricted to these simple vector operations. Over the years, however, a range of collective operations (such as reductions and prefix scans) have been developed that greatly expand the scope of the SIMD pattern. These high-level operations have been incorporated into data parallel languages such as NESL to greatly expand the scope of the SIMD pattern [Blelloch96]. The key point for our discussion is to appreciate that a program using a SIMD pattern lets a programmer understand the concurrency in terms of a single stream of instructions operating on multiple data elements. With the right set of primitives for collective operations, this makes it possible to create deterministic programming models that support the SIMD design pattern. Given the challenge of debugging parallel programs, this determinism can have a powerful impact on the software development cycle.

Known Uses   SIMD algorithms are common on any system that includes a vector processing unit. Often a compiler attempts to extract SIMD vector instructions from a serial program. gRAPHICS processing units traditionally supported the SIMD pattern, but increasingly over time, this is being replaced with closely related but more flexible SPMD patterns. In particular, an SPMD program that makes restricted use of branch statements could be mapped by a compiler onto the lanes of a vector unit to turn a SPMD program into an SIMD program.

Related Patterns   This pattern is a natural choice for many data parallel programs. With a single stream of instructions, it is possible to build SIMD platforms that are deterministic. This can radically simplify the software development effort and reduce validation costs.

As mentioned earlier, programs written using the SPMD pattern can often be converted into an execution pathway that maps onto the SIMD pattern. This is commonly done on graphics processors, and as graphics processors and CPUs become more general and encroach on each other's turf, this overlap between SIMD and SPMD will become increasingly important.

3.3.2.3   Loop-Level Parallelism

Concurrency is expressed as iterations of a loop that execute at the same time.

Use When   This pattern is used for any problem where the crux of the computation is expressed as a manageable number of compute intensive loops. We say "manageable" since a programmer will need to potentially restructure the individual loops to expose concurrency; hence, it is best if the program spends the bulk of its time in a small number of loops.

Details   This pattern is deceptively simple. All you do is find the compute intensive loops and direct the system to execute the iterations in parallel. In the simplest cases where the loop iterations are truly independent, this pattern is indeed simple. But in practice, it's usually more complicated.

The approach used with this pattern is to

- Locate compute intensive loops by inspection or with support from program profiling tools.

- Manage dependencies between loop iterations so the loop iterations can safely execute concurrently in any order.

- Tell the parallel programming environment which loops to run in parallel. In many cases, you may need to tell the system how to schedule loop iterations at runtime.

Typically, the greatest challenge when working with this pattern is to manage dependencies between loop iterations. Of course, in the easiest case, there are no dependencies, and this issue can be skipped. But in most cases, the code will need to be transformed. The following code fragment has three common examples of dependencies in a loop:

```
minval = Largest_negative_int;
for (i=0, ind=1; i<N; i++){
    x = (a[ind+1] + b[ind - 1])/2;
    sum += Z[(int)x];
    if (is_prime(ind)) minval = min(minval,x);
    ind += 2;
}
```

Privatize temporary variables: the variable "x" is a temporary variable with respect to any given iteration of a loop. If you make sure that each thread has its own copy of this variable, threads will not conflict with each other. Common ways to give each thread its own copy of a variable include

- Promotion to an array: Create an array indexed by the thread ID and use that ID to select a copy of the variable in question that is private to a thread.

- Use a built in mechanism provided with the programming environment (such as a private clause in OpenMP) to generate a copy of the variable for each thread.

Remove induction variables: The variable "ind" is used to control the specific elements accessed in the arrays. The single variable is visible to each thread and therefore creates a dependence carried between loop iterations. We can remove this dependence by replacing the variable with an expression computed from the loop index. For example, in our code fragment, the variable ind ranges through the set of odd integers. We can represent this as a function of the loop index

```
ind = 2 * i + 1
```

Reductions: The expression to sum together a subset of the elements of the Z array is called a reduction. The name reduction indicates that a higher dimension object (e.g., an array) is used to create a lower dimension object (e.g., a scalar) through an associative accumulation operation. Most programming models include a reduction primitive since they are so common in parallel programs.

Protect shared variables: In some cases, there are shared variables that cannot be removed. There is no option other than to operate on them as shared variables. In these cases, the programmer must assure that only one thread at a time accesses the variables, i.e., access by any thread excludes any other thread from access to the variable until the variable is released. This is known as mutual exclusion.

Putting these techniques together using OpenMP, we produce the following version of the above loop with the dependencies removed:

```
minval = Largest_negative_int;
#pragma omp parallel for private(x, ind) reduction(sum:+)
for (i=0, ind=1; i<N; i++){
    ind = 2 * i + 1;
    x = (a[ind+1] + b[ind - 1])/2;
    sum += Z[(int)x];
    if (is_prime(ind)){
      #pragma omp critical
        minval = min(minval,x);
}
```

OpenMP is described in more detail in Chapter 6 of this book. We will discuss enough to explain this program fragment. The `pragma` before the loop tells the OpenMP system to fork a number of threads and to divide loop iterations between them. The `private` clause tells the system to create a separate copy of the variables x and ind for each thread. This removes any dependencies due to these temporary variables. The `reduction` clause tells the system to create a separate copy of the variable "sum" for each thread, carry out the accumulation operation into that local copy, and then at the end of the loop, combine the "per-thread" copy of "sum" into the single global copy of sum. Finally, the critical section protects the update to "minval" with only one thread at a time being allowed to execute the statement following the `critical` pragma.

The final step in the "loop parallelism" pattern is to schedule the iterations of the loop onto the threads. In the aforementioned example, we let the runtime system choose a schedule. But in other cases, the programmer may want to explicitly control how iterations are blocked together and scheduled for execution. This is done with a schedule clause in OpenMP, the details of which are described later in the OpenMP chapter.

Known Uses The loop-level parallelism pattern is used extensively by OpenMP programmers in scientific computing and in image processing problems. In both cases, the problem is represented in terms of a grid and updates to the grid are carried out based on a neighborhood of grid points. This maps directly onto nested loops over the grid points which can be run in parallel.

Related Patterns This pattern is used to implement geometric decomposition and data parallel algorithms. When task definitions map onto loop iterations, it is used with task parallel algorithms as well.

Solutions that utilize this pattern may use the SPMD pattern to explicitly parallelize the loops. This approach gives the programmer more control over how collections of loops are divided among threads.

3.3.2.4 Fork-Join

Threads are forked when needed, complete their work, and then join back with a parent thread. The program execution profile unfolds as a series of serial and concurrent phases, often with nesting as forked-threads themselves fork additional threads.

Use When This pattern is used in shared address space environments, where the cost of forking a thread is relatively inexpensive. This pattern is particularly useful for recursive algorithms or any problems composed of a mixture of concurrent and serial phases.

Details In Figure 3.2, we provide a high-level overview of the fork join pattern. Think of the program as starting with a single, serial thread. At points where concurrent tasks are needed, launch or *fork* additional thread to execute a task. These threads

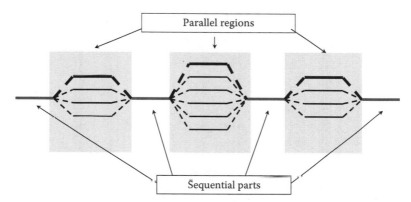

FIGURE 3.2: The fork-join pattern as commonly used with programming models such as OpenMP. Parallelism is inserted as needed incrementally leading to a program consisting of sequential parts (one thread) and parallel regions (teams of threads).

may execute asynchronously from the original serial or master thread. At some later point, the master thread pauses and waits for the forked threads to finish their work. This is called a join. The resulting program consists of serial parts and parallel regions with parallelism added incrementally.

A solution that utilizes this pattern must

- Identify tasks to execute concurrently.

- Expose the concurrency and isolate the task in a form that can be assigned to a forked thread.

- Manage data to prevent conflicting access to data by multiple threads (i.e., to prevent race conditions).

- Join threads back together.

Application of this pattern is heavily influenced by the supporting programming model. Programmers using OpenMP as well as users of native threading libraries such as Pthreads or windows threads make heavy use of this pattern. Thread creation in these models, however, is expensive, so the fork-join pattern is used at a coarse level of granularity.

A more illustrative example of the fork-join pattern is with the programming language Cilk [Blumofe95]. The Cilk runtime makes forking "threads" inexpensive, so the fork-join pattern can be applied in a fine grained manner. The following program recursively computes numbers in the Fibonacci sequence (a pedagogically useful example, but one that is known to be suboptimal for this problem):

```
cilk int fib (int n) {
   int a, b;
   if (n < 2) {
     return n;
   } else {
     a = spawn fib (n-1);
     b = spawn fib (n-2);
     sync;
     return a+b;
   }
}
```

The tasks are the calls to the "fib()" function. The function that contains the instructions implementing these tasks is marked by the keyword "cilk" to indicate to the compiler that this function may be executed concurrently. Inside the function, additional tasks are forked (or in Cilk parlance, spawned) to compute different intermediate results. The "sync" statement in Cilk carries out the function of the join, i.e., it causes the "forking" thread to wait until all "forked" threads complete before proceeding.

Known Uses Every program that uses OpenMP or a native threads library such as Pthreads uses this pattern. The explicit forks and joins, however, may be hidden inside a higher level API. This pattern is particularly important for any program that utilizes recursion. Hence, this is an essential pattern for graph algorithms. The style of programming associated with Cilk also makes heavy use of this pattern.

Related Patterns This pattern is very general and hence it is used to support most of the higher level algorithms patterns. It is particularly important for recursive splitting algorithms.

3.3.2.5 Master-Worker/Task-Queue

A master process defines a collection of independent tasks that are dynamically scheduled for execution by one or more workers.

Use When The master-worker pattern is used for problems that are expressed as a collection of tasks that are independent or that can be transformed into a form where they are independent. In this case, the challenge is to schedule the tasks so the computational load is evenly distributed among a collection of processing elements.

Details A master organizes the parallel computation around a queue of tasks. A set of tasks are created either by a single "master" process or collectively by a set of workers (threads or processes). These tasks are placed in a queue for later processing. Each member from a set of worker threads pulls a task from the queue and does the indicated computation. The results are accumulated either locally on the worker or in more complex cases, in cooperation with a separate master process. When a worker finishes one task, it goes back to the queue for another task. This continues until all the tasks have been handled. Notice that faster workers will naturally take on more tasks. Or if a worker happens to grab more compute intensive tasks, other workers will pick up the load and grab extra tasks. In other words, this approach automatically balances the load of computing the tasks among the set of workers.

Known Uses This is the pattern of choice for embarrassingly parallel problems where the initial problem definition is expressed in terms of independent tasks. Batch queue environments used to share a computational resource among many users are based on this pattern.

Related Patterns The master-worker pattern is frequently used with the task parallel pattern. It is an especially effective pattern to use with an embarrassingly parallel problem since a good implementation of the master-worker pattern will automatically and dynamically balance the load among the workers.

3.4 Conclusions and Next Steps

The patterns in this chapter are just a summary of some of the most important patterns used in parallel programming. This is not a complete set, and there are numerous details we did not include that the complete text of a pattern would include. These pattern summaries, however, should be enough to give you the flavor for what patterns are and how they work. Hopefully, this presentation of patterns will provide valuable insights to you as you read the other chapters in this book.

Experienced parallel programmers reading this discussion of patterns are bound to notice how familiar these patterns seem. This is by design. Patterns organize established knowledge in a domain. They are supposed to capture proven design ideas, not pioneer entirely new design ideas. Hence, we think of the process of defining patterns as "mining" them from excellent solutions to existing problems.

Pattern mining is an ongoing process carried out by a community of "experts" interested in documenting excellence in their field. This is by its nature a work in progress. A pattern language of parallel programming will never be "done" as software designers will inevitably think of new solutions to existing problems or as new classes of problems are addressed. You can follow this ongoing work at [OPL].

References

[Alexander77] C. Alexander, S. Ishikawa, and M. Silverstein, *A Pattern Language: Towns, Buildings, Construction*, Oxford University Press, New York, 1977.

[Blelloch96] G. Blelloch, Programming parallel algorithms, *Communications of the ACM*, 39, 85–97, 1996.

[Blumofe95] R. D. Blumofe, C. F. Joerg, B. C. Kuszmaul, C. E. Leiserson, K. H. Randall, and Y. Zhou, Cilk: An efficient multithreaded runtime system, in *Proceedings of the Fifth ACM SIGPLAN Symposium on Principles and Practice of Parallel Programming (PPoPP)*, Santa Barbara, CA, pp. 207–216, 1995.

[Dean04] J. Dean and S. Ghemawat, MapReduce: Simplified data processing on large clusters, in *Proceedings of OSDI'04: 6th Symposium on Operating System Design and Implementation,* San Francisco, CA, December 2004.

[Gamma94] E. Gamma, R. Helm, R. Johnson, and J. Vlissides, *Design Patterns: Elements of Reusable Object Oriented Software*, Addison-Wesley, Reading, MA, 1994.

[Hwu08] W.-M. Hwu, K. Keutzer, and T. Mattson, The concurrency challenge, *IEEE Design and Test*, 25(4), 312–320, 2008.

[Keutzer10]   K. Keutzer and T. G. Mattson, A design pattern language for engineering (parallel) software, *Intel Technology Journal*, 13(4), 2010.

[Mattson04]   T. G. Mattson, B. A. Sanders, and B. L. Massingill, *Patterns for Parallel Programming*, Addison Wesley, Reading, MA, 2004.

[MPI]   http://www.mpi-forum.org/

[39]   http://www.khronos.org/opencl/

[OPL]   http://parlab.eecs.berkeley.edu/wiki/patterns

[Parsec08]   C. Bienia, S. Kumar, J. Pal Singh, and K. Li, The PARSEC benchmark suite: Characterization and architectural implications, in *Conference on Parallel Architectures and Compilation Techniques*, Toronto, Ontario, Canada, 2008. Available at http://parsec.cs.princeton.edu/overview.htm

[Shaw95]   M. Shaw and D. Garlan, *Software Architecture: Perspectives on an Emerging Discipline*, Prentice Hall, Upper Saddle River, NJ, 1995.

Part II

Programming Languages for Multicore

Chapter 4

Threads and Shared Variables in C++

Hans Boehm

Contents

As of 2009, neither C nor C++ directly supported parallel programming in the language itself. Traditionally, C++ parallel programs most commonly relied on platform-dependent libraries to allow the creation of multiple threads of control and to support synchronization between them. The two most common interfaces for such libraries are the Posix "pthreads" interface [9], and Microsoft Windows' threads interface.

Unfortunately these approaches resulted in some ambiguity and confusion about the meaning of threads, and the properties that compilers were expected to enforce with respect to threads [4]. As a result, it was difficult to precisely define the resulting programming rules.

The C++ language is currently undergoing revision. This upcoming version of the language is customarily referred to as C++0x, in spite of the fact that it is not expected to be finalized until roughly 2011.

C++0x addresses the difficulties with threads in C++ by adding support for threads directly to the language, giving the language specification a chance to precisely define their semantics [6]. The next revision of the C standard is taking a similar route though with significant differences in the API for thread creation and synchronization.

In this chapter, we describe multi-threaded programming in C++0x, since it significantly simplifies our task. Aside from significant syntactic differences, the approach is fundamentally similar to C++03 with the Boost threads [16] library. However, the basic rules are clearer, simpler, and much less platform-dependent.

4.1 Basic Model and Thread Creation

C++0x threads each independently execute their own program, but they do so in a completely shared address space. Variables may be statically allocated to fixed locations, by declaring them outside of any function, or with a `static` qualifier, they may be local to a particular function activation, or they may be declared to be local to a thread with the `thread_local` qualifier. In the latter two cases, they are directly accessible to only a single thread. However, that thread may store a pointer to such a variable l in a statically allocated variable p, and thus allow other threads to refer to it as $*p$.

Fundamentally, all areas of memory are accessible from all threads. As in sequential C++, it is the programmer's responsibility to ensure that variables are not accessed beyond their lifetimes. That clearly requires some care when one thread accesses variables intended to be local to another thread, but we will see later that it can be both safe and useful.

In practice, there are also C and C++ programs that use completely different approaches to parallelism through the use of other libraries, or by relying directly on OS facilities. It is fairly common to write parallel C++ applications that either share no memory at all and communicate via some form of message passing, or that communicate via shared memory regions, but do not share the entire process address space. Both have their advantages and disadvantages. The latter can be significantly more complicated than the threads-based model, since the programmer has to track which addresses are valid in which process. Neither of these are directly supported by the C++0x draft standard, and we do not directly address them in this chapter.

Each C++0x thread has an associated object of type `std::thread`. A thread is created and started by constructing an object of type `std::thread` with arguments specifying the function to be invoked in its thread, together with any necessary arguments. Thus, we could start a thread computing `sqrt(1.23)` by declaring a thread object `t`:

```
std::thread t(sqrt, 1.23);
```

However, such a thread would not be useful, since the result of the sqrt function is discarded.

Once the thread object t has been created, it is the creator's responsibility to ensure that t.join() is subsequently invoked to wait for the thread and allow its resources to be reclaimed. If join() is not called, a run-time error results when t is destroyed, for example, because it goes out of scope.

Invoking t.join() causes the invoking thread to wait for the thread represented by t to complete. This is usually necessary so that we can safely use the results it produced. Even if we use other mechanisms to wait for results, it is typically necessary to wait for thread completion to ensure that the thread terminates before resources it relies on are destroyed. We address this point in more detail in Section 4.10.

4.2   Small Detour: C++0x Lambda Expressions

In order to simplify examples, and give our first illustration of a (minimally) useful thread, we briefly introduce another C++0x feature, so-called lambda expressions. These provide a convenient syntax for constructing objects that behave like functions, but can refer to variables in the surrounding scope. (Unlike its predecessors in other languages, the syntax does not use the Greek letter λ; the name still derives from Church's lambda calculus.)

The expression

```
[&]() { x = sqrt(1.23); }
```

denotes a function object, that is, an object that can be invoked as a function. If x is declared in a surrounding function, then it is bound by reference (indicated by the & inside the square brackets introducing the function) to that instance in the surrounding scope.

If the brackets had contained an "=" sign, variables like x that are neither local to the lambda expression nor completely global would have been *copied* into the function object, rather than accessed by reference. This would be completely useless here, since we want to change the value of x rather than examining its initial value. But it is often useful in other contexts. The full syntax also allows finer control of how individual variables are captured (by reference or value), and the leading "&" or "=" specifies the default treatment.

4.3   Complete Example

Using this notation, we can now write a function that actually does compute sqrt(1.23) in a separate thread, though clearly this would not be useful unless sqrt(1.23) took exceptionally long to compute, and we had something else to do in the interim:

```
#include <thread>

double sqrt123()
{
  double x;
  std::thread t([&]() { x = sqrt(1.23); });
  // parent thread might do something else here.
  t.join();
  return x;
}
```

The function invoked by the thread t first computes sqrt(1.23) and then assigns the result to the variable x in the creating (parent) thread. The parent thread then waits for t to finish before returning the final value of x.

Note that if the parent thread throws an exception between the creation of t and the call to t.join(), the program will terminate since t will be destroyed without an intervening t.join() call. If there is danger of such a recoverable exception, the code should ensure that t.join() is also called on the exception path.

4.4   Shared Variables

As in other programming languages supporting threads, C++ threads communicate by accessing shared variables. We normally think of the threads as executing in an interleaved fashion, with one thread executing one or more steps, followed by another thread executing some number of steps, etc. Thus, the function in Figure 4.1 might be executed as

Main Thread	Child Thread

`int x = 0;`
`create child;`
`x = 200;`	

	`x = 400;`
join child; `return x;` // returns 400	

```
int f()
{
  int x = 0;
  std::thread t([&]() { x = 400; });
  x = 200;
  t.join();
  return x;
}
```

FIGURE 4.1:   Simultaneous assignment.

or as

Main Thread	Child Thread
`int x = 0;` `create child;`	
	`x = 400;`
`x = 200;` `join child;` `return x;` // returns 200	

Such an execution is called *sequentially consistent* [12] if it can be understood in this way as the interleaving of thread steps. In a sequentially consistent execution, the value of a shared object is always the last value assigned to it in the interleaving, no matter which thread performed the assignment.

Although this is the traditional way to view the execution of a multi-threaded program, it is generally viewed as impractical, especially for a systems programming language, such as C++, for two different but complementary reasons:

1. The meaning of a program depends on the *granularity* at which memory operations are performed. To see this, consider a hypothetical machine on which memory accesses are performed a byte at a time. Thus, the assignment `x = 400` is actually decomposed into at least the two assignments `x_high_bits = 1` and `x_low_bits = 144` ($1 \times 2^8 + 144 = 400$). Thus, we could also get an interleaving

Main Thread	Child Thread
`int x = 0;` *create child;*	
	`x_low_bits = 144;`
`x_low_bits = 200;`	
	`x_high_bits = 1;`
join child; `return x;` // returns $256 + 200 = 456$	

This result is almost certainly unexpected. Even if the hardware performs memory accesses at the expected granularity, operations on user-defined data structures are unlikely to be performed indivisibly, and thus would have to be thought about differently from operations on, say, integers. Although sequential consistency is probably the easiest model to understand, it is often not very helpful for practical reasoning.

2. Both the compiler and hardware routinely perform optimizations that violate sequential consistency, in that they produce results that cannot be understood in such a simple interleaving model.

   Consider the example in Figure 4.2, where we assume that individual memory accesses are indivisible. Based on the interleaving interpretation, either `x = 1` or `y = 1` must be executed first. Thus, the corresponding load in the other thread must return one, and thus either `res_parent` or `res_child` must be nonzero, and `f()` can never return true.

```
bool f()
{
  int x = 0, y = 0;
  int res_parent, res_child;
  std::thread child([&]() { x = 1; res_child = y;});
  y = 1;
  res_parent = x;
  child.join();
  return res_parent == 0 && res_child == 0;
}
```

FIGURE 4.2:   Can f() return 0?

However, nearly all modern implementations do allow f() to return true. Compilers may decide to reorder the memory operations in each thread, performing the load from the shared variable x or y first, allowing, for example, the interleaving

  res_parent = x; x = 1; res_child = y; y = 1;

(it is often attractive to perform loads earlier, since it may give the load operation more time to complete before the result is needed). Even if the compiler does not, typical hardware performs a similar transformation by buffering the initial store to x or y, making it visible to other processors later, possibly after the subsequent load from the other shared variable has completed. In either case, both shared variable loads can yield zero.

4.5   More Refined Approach

Since sequential consistency is neither sufficient to avoid surprising outcomes and thus ensure a convenient programming model, nor reliably implementable with good performance, C++0x instead follows Java, and adopts a slightly different programming model that guarantees sequential consistency only for a class of well-behaved programs that excludes the preceding example [1,2,6,13].

The problem in all of the preceding cases is that shared variables are simultaneously accessed by two different threads. This allows one thread to observe whether operations in the other thread are reordered or performed with small granularity. We avoid these issues completely by prohibiting such simultaneous accesses, commonly referred to as *data races*.

More precisely, C++0x defines a *memory location* to be either a scalar (e.g., int, char, double, or pointer) or a sequence of contiguous bit-fields.* This is illustrated in Figure 4.3 (more on bit-fields below.) Informally, a memory location is a unit of

---

* As a special exception, zero-length bit-fields can be used to separate sequences of bit-fields into multiple memory locations.

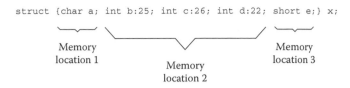

FIGURE 4.3: Memory locations in a `struct`.

memory that can be updated without also writing to adjacent memory, which might interfere with other threads.

Two memory operations `conflict` if they access the same memory location and at least one of them is a store operation to that location. A particular execution of a program (interleaving) contains a *data race* if two conflicting operations corresponding to different threads occur adjacently in the interleaving, reflecting the fact that they are not prevented from executing concurrently. For example, Figure 4.1 has a data race because the two assignments to x conflict, and can be executed simultaneously by different threads, as is seen by the fact that they can be adjacent in an interleaving. Similarly, Figure 4.2 contains two distinct data races: The assignment and access to x conflict and may be simultaneously executed, and similarly for y.

A program with a possible execution that includes a data race has "undefined behavior," that is, allowed to produce any results whatsoever. (This is sometimes referred to as "catch fire" semantics for data races, though we do not expect real implementations to actually cause a fire.) A data race is erroneous, though it is difficult to diagnose such errors, and typical implementations generally will not. Typical implementations will, however, perform program transformations that assume there are no data races. When that assumption is violated, very strange behavior may result.

To understand the special treatment for bit-fields in the definition of "memory location," consider how they are typically implemented. Assume x is a `struct` consisting of two 4 bit wide fields a and b sharing a single byte. Essentially no modern hardware allows either field to be changed without storing to the entire byte. Thus, `x.a = 1` would be implemented as

```
tmp = x; tmp.a = 1; x = tmp;
```

where tmp is stored in a register, and the middle assignment involves some bit manipulation instructions operating on the register value. This means that if both fields are initially zero, and two threads concurrently assign to the two fields, we could get the following interleaving of the generated code:

Thread 1	Thread 2
`tmp = x;`	
	`tmp2 = x;`
	`tmp2.b = 1;`
`tmp.a = 1;`	
`x = tmp;`	
	`x = tmp2;`

This resulting code clearly has a data race on x, and this is reflected in the unexpected result here, in which only the x.b field is actually updated, since the second thread happens to write last. In order to avoid the introduction of such data races, the draft standard treats the original code as updating the same memory location in both threads, and hence as already containing the data race.

By effectively outlawing data races, we ensure that legal code cannot test whether memory operations are reordered and cannot detect the granularity at which they are performed. Even larger operations, such as operations on STL containers, can be treated as though they occur in one atomic operation; any code that could tell the difference is already outlawed by the preceding rule.

4.6 Avoiding Data Races

Unfortunately, merely outlawing data races is not enough; so far, we have no clear mechanism to avoid data races if threads access the same shared variable, as most interesting programs will want to.

Consider a simple program consisting of several threads, which, as part of their operation, need to very occasionally increment a shared counter c. If we incremented the counter by simply executing c++, we would introduce a potential data race. Each of the increment operations consists of both a load and a store operation of the variable c. There is nothing to prevent operations performed by two threads from being executed concurrently or, more precisely, from a load operation on c from thread *a* appearing next to a store operation on c from thread *b*.

C++ provides two primary synchronization mechanisms for avoiding data races in cases like this:

Mutexes: Mutexes prevent certain code sections from being interleaved with each other. Informally, a mutex (class std::mutex) cannot be *acquired* by a thread, by invoking lock() function on the mutex, until all prior holders of the mutex have released it by calling its unlock() method. More formally, a lock() call is not allowed to occur in the interleaving until the number of prior unlock() calls on the mutex is equal to the number of prior lock() calls.

Atomic objects: Operations on atomic objects appear to occur indivisibly. They do not participate in data races; the real definition of data races in the draft standard refers only to ordinary memory accesses, specifically excluding atomic accesses.

4.7 Mutexes

To implement the shared counter using the preceding locks, we might write the following:

```
#include <thread>
#include <mutex>
using namespace std;
mutex m;
int c;

void incr_c()
{
  m.lock();
  c++;
  m.unlock();
}
```

Although this is occasionally the right coding style for a code that acquires locks, it introduces a major trap: If the code between the lock() and unlock() calls throws an exception, the lock will not be released. This prevents any other thread from acquiring the lock, and would thus normally cause the program to deadlock in short order.

One could avoid this by explicitly catching exceptions and also unlocking on the exception path. But the standard provides a more convenient mechanism: A std::lock_guard object acquires a mutex when it is constructed, and releases it on destruction. Since the destructor is also invoked in any exception path, the underlying mutex will be released even when an exception is thrown.

Lock_guard uses the standard C++ RAII (resource acquisition is initialization) idiom. Like some other uses of this idiom, the actual lock_guard variable is not of interest. (We give it a name of "_" to emphasize that.) Its only purpose is to ensure execution of its constructor and destructor. Thus, the earlier example could be rewritten as follows:

```
#include <thread>
#include <mutex>
using namespace std;
mutex m;
int c;
void incr_c()
{
  lock_guard<mutex> _(m);
  c++;
}
```

4.8 Atomic Variables

C++0x provides an alternate synchronization mechanism that may be more appropriate for very simple shared data, like our counter earlier. By declaring a variable to have type `atomic<T>` instead of T, accesses to it are performed indivisibly, and pairs of these accesses cannot constitute data races. The implementation is required to ensure that the resulting programs still behave sequentially consistently, that is, as if steps from different threads were interleaved. Depending on the platform, it may make accesses to `atomic<T>` objects much more expensive than regular accesses to T.

The type `atomic<T>` may be used for any T that can be copied by simply copying the bits. It always provides `load()` and `store()` operations. The latter is also provided as an assignment operator, and the former is also provided as an explicit conversion from `atomic<T>` to T; thus, usually, neither `load()` nor `store()` operations are written explicitly. In addition, a few atomic read-modify-write operations are provided. In particular, increment operations on `atomic<int>` are indivisible.

Thus, the example from the preceding section could be written more concisely, and usually more efficiently, as follows:

```
#include <thread>
#include <atomic>
using namespace std;
atomic<int> c;

void incr_c()
{
    c++;
}
```

Although atomic operations by default continue to provide sequential consistency for data-race-free programs, there are also versions of these operations, customarily referred to as "low-level" atomics, that violate the simple interleaving semantics for improved performance. Unfortunately, on current hardware, these performance improvements can sometimes be dramatic.

The semantics of low-level atomics are quite complex, and the details are beyond the scope of this chapter.

We expect that code using `atomic` objects will generally be written initially using "high-level" sequentially consistent atomics, and hot paths may then be carefully manually optimized to use low-level atomics, where it is essential to do so. By doing so, we at least separate subtle memory ordering concerns from the already sufficiently subtle issue of developing the underlying parallel algorithm. We briefly illustrate this process here, describing the simplest and probably most common application of low-level atomics in the process.

Assume we would like to initialize some static data on the first execution of a function. (We could also do this by simply declaring a function-local static variable, or the standard-library-provided `call_once` function, in which case the language implementation might itself use a technique similar to the one described here.)

A simple way to do so explicitly would be to write the appropriate code using locks as shown earlier. To implement a function `get_x()` that returns a reference to x, which is guaranteed to have previously been explicitly initialized, we might write the following:

```
#include <mutex>
#include <atomic>
using namespace std;

T x;
bool x_init(false);
mutex x_init_m;

T& get_x()
{
  lock_guard<mutex> _(x_init_m);
  if (!x_init) {
    initialize_x();
    x_init = true;
  }
  return x;
}
```

This has the, often substantial, disadvantage that every call to get_x() involves the overhead of acquiring and releasing a lock. We can improve matters by using the so-called double-checked locking idiom:

```
T x;
atomic<bool> x_init(false);
mutex x_init_m;

T& get_x()
{
  if (!x_init) {
    lock_guard<mutex> _(x_init_m);
    if (!x_init) {
      initialize_x();
      x_init = true;
    }
  }
  return x;
}
```

We first check whether x has already been initialized *without* acquiring a lock. If it has been initialized, there is no need to acquire the lock.

Note that in this version, x_init can be set to true in one thread while being read by another. This would constitute a data race had we not declared x_init as atomic. Indeed, in the absence of such a declaration, if the compiler knew that the implementation of initialize_x did not mention x_init, it would have been perfectly justified in, for example, moving the assignment to x_init to before the initialization of x, thus effectively breaking the code.

By declaring x_init to be atomic, we avoid these issues. However, we now require the compiler to restrict optimizations so that sequential consistency is preserved and, more importantly, to insert additional instructions, typically so-called memory fences, that prevent the hardware from violating sequential consistency.

4.8.1   Low-Level Atomics

On most existing architectures, this version of the code is likely to perform adequately. The assignment to x_init will typically be more expensive than necessary, but it is only executed once. Unfortunately, on some architectures, the other accesses to x_init will also incur additional overhead [14].

If we want to reduce this additional overhead, and were feeling exceptionally brave, we could do so by explicitly specifying ordering constraints for the atomic operations, and sacrificing the simple interleaving model of thread execution, at least around this piece of code. We could now rewrite the code as follows:

```
T x;
atomic<bool> x_init(false);
mutex x_init_m;

T& get_x()
{
  if (!x_init.load(memory_order_acquire)) {
    lock_guard<mutex> _(x_init_m);
    if (!x_init.load(memory_order_relaxed)) {
      initialize_x();
      x_init.store(true, memory_order_release);
    }
  }
  return x;
}
```

The memory_order_release store to x_init ensures that all memory accesses preceding it become visible to any other thread after it performs a memory_order_acquire load of x_init that sees the new value. This is the only visibility guarantee required here for the concurrent accesses to x_init. For example, it prevents the compiler from moving up the store to x_init.

The combination of memory_order_release and memory_order_ acquire does not ensure sequential consistency. Most importantly, it does not guarantee that atomic stores followed atomic loads from a different variable become visible to other threads in the intended order.

Assume two threads execute the following code, where both x and y are initially zero. (This is essentially similar to the last example in Section 4.4, only presented in a different format. This is commonly referred to as the "Dekker's" example, since it is the core of Dekker's mutual exclusion algorithm.)

Thread 1
`x.store(1, memory_order_release);` `res_1 = y.load(memory_order_acquire);`
Thread 2
`y.store(1, memory_order_release);` `res_2 = x.load(memory_order_acquire);`

It would be entirely possible to get a final result of `res_1 = res_2 = 0`, since the loads could appear to be reordered with the stores.

For our specific `get_x()` example, this kind of reordering is typically not an issue, so the faster primitives are safe for this idiom. It appears to be the case that no caller of `get_x()` can tell whether the faster primitives are used, and hence the caller may continue to reason based on interleavings, that is, sequential consistency. But even in this simple case, we know of no rigorous proof of that claim. In general, it appears quite difficult to effectively hide the use of low-level atomics inside libraries, keeping them invisible from the caller.

Note that the second load from `x_init` in `get_x()` cannot in fact occur at the same time as a store to `x_init`. Thus, it is safe to require no ordering guarantees. One perfectly safe use of low-level atomics is for such non-racing accesses to variables that need to be declared `atomic` because other accesses are involved in races.

4.9 Other Synchronization Mechanisms

In addition to atomic types and locks, C++0x provides a fairly standard set of additional synchronization facilities. We briefly list here the major ones.

4.9.1 Unique_lock

The `lock_guard` facility provides a useful facility for managing the ownership of mutexes, and ensuring that mutexes are released in the event of an exception. However, it imposes one important restriction: The underlying mutex is unconditionally released by the `lock_guard`'s destructor. This means that the mutex can never be released within the scope of the `lock_guard`.

The `unique_lock` template provides a generalization of `lock_guard` that supports explicit locking and unlocking within the scope of the `unique_lock`. The `unique_lock` tracks the state of the mutex, and the destructor unlocks the mutex only if it is actually held. In this way, the programmer can get essentially the full flexibility of direct mutex operations, while retaining the exception safety provided by `lock_guard`.

4.9.2   Condition Variables

Like most threads implementations since Mesa [15], C++ provides *condition variables*.

A thread may need to wait for some condition to hold in order to make progress. For example, we may have a thread whose function it is to look at a queue of pending requests, and to process those requests as it finds them.

Once the queue becomes empty, it could wait for it to fill again by repeatedly testing an `atomic` variable containing, say, the number of elements in the queue. This is clearly wasteful, since the thread will continue to consume CPU resources, thus possibly slowing down the thread refilling the queue, while it is waiting.

A better solution is to associate a condition variable with the "queue nonempty" condition. This might be declared as follows:

```
#include <condition_variable>

std::condition_variable queue_nonempty;
```

Assume that q is the shared `queue`. Since it is shared, it must be protected by a mutex `q_mtx`, and all accesses to q must occur by a thread holding `q_mtx`.

In order to test q, we must hold `q_mtx`. We do so by declaring a suitable `unique_lock`. If we discover that q is empty, we then wait for the condition variable to be *notified*, indicating that something was added to the queue, and it might hence be nonempty. Thus, the code to inspect the queue might look something like

```
{
   std::unique_lock<std::mutex> q_ul(q_mtx);

   while (q.empty()) {
      queue_nonempty.wait(q_ul);
   }
   // retrieve element from q
}
```

Note that it makes no sense to wait for the queue to be refilled while holding `q_mtx`, since that would make it impossible for another thread to add anything to the queue while we are waiting. Hence, the `wait()` call on a `condition_variable` needs a way to release the mutex while it is waiting. This is done by passing it a `unique_lock`, which unlike a `lock_guard` can be released and required. (One might alternatively have designed the interface to pass a raw `std::mutex`. That would have made it much more difficult to ensure proper release of the mutex in case of an exception.)

When another thread adds something to the queue, it should subsequently invoke either

```
queue_nonempty.notify_one();
```

or

```
queue_nonempty.notify_all();
```

The former allows exactly one waiting thread to continue. It is normally more efficient in cases like our example in which there is no danger of waking the wrong thread, and thus potentially reaching deadlock. The latter allows all of them to continue, and is generally safer. If there are no waiting threads, neither call has any effect.

Note that `condition_variable::wait()` calls *always* occur in a loop. (The `condition_variable` interface also contains an overloaded `wait` function that expects a `Predicate` argument, and executes the loop internally.) In most realistic cases, for example, if we have more than one thread removing elements from q, this is usually necessary because a third thread may have been scheduled between the notify call and the wait call, and it may have invalidated the associated condition, for example, by removing the just added element from the queue. Since the client code generally needs a loop anyway, the implementation is somewhat simplified by allowing `wait()` to occasionally return *spuriously*, that is, without an associated notification. Thus, the loop is required in all cases.

Since `wait()` can wake up spuriously, the correct code should remain correct if all wait and notify calls are removed. However, this would again be far less efficient. It might conceivably even prevent the application from making any progress at all, if waiting threads consume all processor resources, though this is not likely for mainstream implementations.

The implementation of `wait()` is guaranteed to release the associated mutex atomically with the calling threads entrance into the waiting state. If the condition variable is notified while holding the associated mutex, there is no danger of the notification being lost because the waiting thread has released the mutex, but is not yet waiting, a scenario referred to as a "lost wakeup." Thus, the `wait()` call must release the mutex, and condition variables are tightly connected with mutexes; there is no way to separate `condition_variable::wait()` and the release of the mutex into separate calls.

A C++ condition variable always expects to cooperate with a `unique_lock<mutex>`. Occasionally it is desirable to use condition variables with other types of locks. `Std::condition_variable_any` provides this flexibility.

4.9.3 Other Mutex Variants and Facilities

The basic `mutex` type can only be acquired once per thread. A mutex can only be acquired by a thread *t* if neither any other thread nor *t* itself holds the lock. A `recursive_mutex` may be acquired even if *t* itself already holds the lock. The thread *t* must then release the mutex the same number of times it previously acquired it before another thread can acquire it.

A simple implementation of an object shared between threads would typically include a mutex in the object. Any public functions would then acquire the mutex before operating on the object state, thus ensuring that simultaneous accesses to the object do not result in data races. Private, internal functions would not acquire the mutex, since they would be called from a public function that already owns the mutex. The use of a `recursive_mutex` would allow a public function owning the mutex to call another public function, which would reacquire it. If a regular `mutex` were used in this setting, the second acquisition attempt would either result in deadlock or throw an exception indicating detection of a deadlock condition.

Either kind of mutex can be acquired either by calling its `lock()` function, which blocks, that is, waits until the lock is available, or by calling its `try_lock()` function, which immediately returns `false` if the mutex is not available. The `try_lock()` function may spuriously return `false`, even if the mutex is available. (Implementations will typically not really take advantage of this permission in the obvious way, but this permission declares certain dubious programs to contain data races, and it would potentially be very expensive to provide sequential consistency for those programs. See Ref. [6] for details.)

One use of the `try_lock()` function is to acquire two (or more) locks at the same time, without risking deadlock by potentially acquiring two locks in the opposite order in two threads, and having each thread successfully acquire the first one. We can instead acquire the first lock with `lock()` and try to acquire the second with `try_lock()`. If that fails, we release the first and try the opposite order. In fact, the standard supplies both `lock()` and `try_lock()` stand-alone functions that take multiple mutexes or the like, and follow a process similar to this to acquire them all without the possibility of deadlock, even in the `lock()` case.

The draft standard also provides mutex types `timed_mutex` and `recursive_timed_mutex` that support timeouts on lock acquisition. Since the draft standard generally does not go out of its way to support real-time programming, for example, it does not support thread priorities, these are expected to be used rarely.

4.9.4   Call_once

The `call_once` function is invoked on a `once_flag` and a function f plus its arguments. The first call on a particular `once_flag` invokes f. Subsequent calls simply wait for the initial call to complete. This provides a convenient way to run initialization code exactly once, without reproducing the code from Section 4.8.

As we also mentioned earlier, often an even easier way to do this is to declare a static variable in function scope. If such a variable requires dynamic initialization, it will be initialized exactly once, on first use, using a mechanism similar to `call_once`.

4.10   Terminating a Multi-Threaded C++ Program

All of our example programs have created a thread, and then waited for it to complete using `thread::join()`. In some cases, it is convenient to create one or more threads that provide a service for a particular library, and to essentially keep them running until program termination. Or it may make sense to create a thread to produce some external output that the rest of the program does not logically need to wait for.

Posix [9] provides a `pthread_detach()` call specifically to make it more convenient to manage such threads, and it is tempting to do something similar in C++. However, C++ presents an added difficulty here: When `exit()` is called to terminate the process, typically because the initial thread returns from `main()`, destructors for static duration variables are invoked. If another thread is still running at this time, it is likely to make library calls that touch those same static variables, typically resulting in a potential data race. In practice, such a thread that happens to be running at the wrong time, particularly near the end of static destruction, is likely to access statically allocated objects that have already been destroyed, resulting in a crash or misbehavior during process shutdown.

This issue is somewhat complicated by the fact that, even after a thread returns from the top-level function passed to the thread constructor, it may continue to run destructors for objects declared `thread_local`, making it harder to wait for full completion of a thread, except by joining it.

As a result, there are basically two distinct safe ways to shut down a C++ process:

- The main (original) thread must ensure that all threads have been joined before it returns. This implies that any library that starts a helper thread should either provide an explicit call to shut down those helper threads, and those calls should be invoked before `main()` exits, or it should join with the helper threads when one of the library's static duration objects is destroyed. The latter option is quite tricky, since such a thread may not rely on another library that may be shutting down at the same time. The standard library is guaranteed to be available until the end of static duration object destruction, but user-level libraries typically do not share this property, since their static objects may be destroyed in parallel.

- The entire process may be shut down without running static destructors by calling `quick_exit`. This will simply terminate all other threads still running at that point. It is possible to perform some limited shutdown actions in the event of a `quick_exit` call by registering such actions with `at_quick_exit()`. However, any such registered actions should be limited so that they do not interfere with any threads running through the shutdown process.

Unfortunately, any use of the second option is not fully compatible with existing C++ code that relies on the execution of static destructors. Essentially the entire application has to agree that `quick_exit` will be used, and static destructors will not be used. As a result, we expect most applications to use the first option, and to ensure that all threads are joined before process exit.

4.11   Task-Based Models

Here, we have described only concurrency as supported by the (upcoming) language standard itself. There is general agreement that this is a lower level than should be used by the majority of programmers.

The model we have presented here requires that a separate thread be created for tasks to be performed in parallel. If this is done indiscriminately, it can easily result in more concurrently active threads than the runtime can efficiently support, and in far more thread creation and destruction overhead than desired, possibly outweighing any benefit from performing the tasks in parallel in the first place. Essentially, we have left it to the programmer to divide an algorithm into tasks of sufficiently small granularity to keep all available processors busy, but sufficiently large granularity to prevent thread creation and scheduling overheads from dominating the execution time.

A variety of library facilities have been proposed to mitigate these issues:

- Parallel algorithms libraries (cf. STAPL [3]) that directly provide for parallel iteration over data structures, and thus the library either simplifies or takes responsibility for the choice of task granularity. These often also support containers distributed over multiple machines, an issue we have not otherwise addressed here.

- Thread pools allow a small number of threads to be reused to perform a large number of tasks.

- Fork-join frameworks such as Intel Threading Building Blocks [10,11] or Cilk++ [7,8] allow the programmer to create a large number of potentially parallel tasks, letting the runtime make decisions about whether to run each one in parallel, essentially in a thread pool, or to simply execute it as a function call in the calling thread. This allows the programmer to concentrate largely on exposing as much parallelism as possible, leaving the issue of limiting parallelism to minimize overhead to the runtime.

Particularly, the latter two often allow the cost of creating a logically parallel task to be reduced far below typical thread creation costs. The next chapter discusses some related approaches in other languages in more detail.

Although such facilities were discussed in the C++ committee, none were incorporated into C++0x. Such facilities are quite difficult to define precisely and correctly. Issues like the following need to be addressed:

- How do `thread_local` variables interact with tasks? Does a task always see consistent values? Can they live longer than expected, leaking memory, and potentially outliving some data structures required to execute the destructor?

- What synchronization operations are the tasks allowed to use? What about libraries that use synchronization "under the covers"? Usually, serializing execution of logically parallel and unrestricted tasks introduces deadlock possibilities. What rules should the programmer follow to avoid this?

Although there is agreement that we would like to encourage such programming practice, the prevailing opinion was that it was premature to standardize such features without a deeper understanding of the consequences, particularly in the context of C++.

C++0x does include some minimal support targeting such approaches, which were finalized late in the process:

- Class promise provides a facility for setting and waiting for a result produced by a concurrently executing function. This provides a set_value member function to supply the result, for example, in a child thread, and a get_future function to obtain a corresponding future object, which can be used to wait for the result, for example, in a parent thread. This allows propagation of exceptions from the child to the parent. Class packaged_task can be used to simplify the normal use case, by wrapping a function so that its result is propagated to an associated future object.

- The function template async allows very simple and convenient execution of parallel tasks. In order to avoid the aforementioned problems, such tasks are always executed either in their own thread or, by default, at the runtime's discretion, sequentially at the point at which the result is required. We could rewrite our first complete threads example as follows:

```
#include <future>

double sqrt123()
{
    auto x = std::async([]() { return sqrt(1.23); });
    // parent thread might do something else here.
    return x.get();
}
```

Here x is a future, which is returned immediately by the async() call. The get() call waits for the result to be ready.

Aside from removing the risks associated with having the child thread write directly to the parent's variable, it ensures that exceptions are correctly propagated.

4.12 Relationship to Earlier Standards

Although we have focused on threads in C++0x, the approach we have described is not fundamentally different from what would be used with C++03 or C99. However, those platforms present some additional challenges. We list those here.

4.12.1 Separate Thread Libraries

A separate thread library must be used. It is probably most common to use a library that exports a C language API, such as the Posix threads API [9] or the Microsoft

Windows API. The former is closer to what we have presented here than the latter. It is also possible to use one of several possible add-on C++ threads APIs, which are usually implemented as a thin layer over the underlying C-level API. The C++0x threads API is very loosely based on one of these, namely, that provided by Boost [16].

4.12.2   No Atomics

There is no full equivalent of C++0x `atomic<T>` variables. There are a number of vendor-specific extensions that provide related functionality, such as Microsoft's `Interlocked...` functions or gcc's `__sync...` intrinsics. However, these provide primarily atomic read-modify-write operations, and it is unclear how to safely implement plain loads and stores on such shared values. They also occasionally provide surprising memory ordering semantics.

Note that it is generally not safe to access variables that may be concurrently modified by another thread using ordinary variable accesses. Such an attempt may have results that do not correspond to reading any specific value of the variable. For example, consider

```
{
    bool my_x = x;   // x is shared variable

    if (my_x) m.lock();
    a: ...
    if (my_x) m.unlock();
}
```

The compiler may initially load `my_x`, use that value for the first conditional, then be forced to "spill" the value of `my_x` when it runs out of available registers, and then reload the value of `my_x` from x for the second condition, since it knows that `my_x` was just a copy of x. If the value of x changes from `false` to `true`, the net result of this is a likely runtime error as `m.unlock()` is called without a prior call to `m.lock()`, something that initially appears impossible based on the source code.

Operations on C++ `volatiles` do put the compiler on notice that the object may be modified asynchronously, and hence are generally safer to use than ordinary variable accesses. In particular, declaring x volatile in the preceding example would prevent `my_x` from being reloaded from it a second time. However, `volatile` does not in general guarantee that the resulting accesses are indivisible; indeed it cannot, since arbitrarily large objects may be declared `volatile`. On most platforms, there are also very weak or no guarantees about memory visibility when `volatile` is used with threads. We know of no platform on which the Dekker's example would yield the same behavior with `volatile int` as it would with `atomic<int>`.

In general, there is no fully portable replacement for `atomic<T>`. The usual solutions are to either use locks instead, to use non-portable solutions, or to use a third-party library that attempts to hide the platform differences.

4.12.3 Adjacent Field Overwrites

In C++0x, an assignment may only affect the "memory location" in which it resides, and "memory locations" are defined such that an assignment to a variable never writes an adjacent variable, and writing to a field may read and rewrite an adjacent field only if both are members of the same contiguous sequence of bit-fields. Thus, for a structure declared as:

```
struct S {char a; int b:5; int c:11; char d;}
```

All fields may be updated in parallel, except that concurrent assignments to b and c may introduce a data race.

Older specifications are generally sufficiently ambiguous about this that we have not found them helpful. However, most recent implementations differ from the C++0x specification primarily in that assignments to bit-fields will read a unit of the structure corresponding to the declared type of the bit-field. Thus, in our example, an update of b or c is likely to read an `int`-sized part of the structure, update that in a register, and write it back. On typical machines with 32 bit `int`s, this typically means that *no* concurrent updates of any of the fields in `struct S` are safe. Although this has major impact for our specific example, this kind of structure with embedded bit-fields is rare in most real code. Putting bit-fields into a separate structure is safer, though may cost some space.

On a few architectures, notably Alpha machines, compiler switches may be required to avoid more surprising overwrites of adjacent fields.

4.12.4 Other Compiler-Introduced Races

Older compilers tend to occasionally introduce a code that amounts to additional identity assignments (i.e., $x = x;$) under certain unusual conditions. This may overwrite an assignment to the same variable (i.e., x) performed by another thread between the load and the store of the identity assignment. Struct field assignments from the preceding section are one example of this. But there are others.

As another example, consider the loop

```
for (...) {
   if (...) ++count;
}
```

where `count` is a global, shared variable. It is fairly common to compile this to the equivalent of:

```
reg = count;
for (...) {
   if (...) ++reg;
}
count = reg;
```

If the loop body is in fact not executed, this effectively introduces a `count = count;` assignment where none was present in the source. This potentially introduces a data race, though only if the programmer happened to know that this loop executed for no iterations, and concurrently accessed `count`. There are other similar, though much less likely, transformations that introduce races into much more plausible code [4].

In practice, programmers rely on the fact that such transformations are very unlikely to introduce problems. We know of no real work-around other than disabling some rather fundamental compiler optimizations that only rarely violate the C++0x rules. Since such transformations are no longer legal in C++0x, we expect them to rapidly disappear, and a few compilers are already careful to avoid them.

4.12.5 Program Termination

Unfortunately, existing multi-threaded C++ programs often overlook the fact that a thread running until process termination is likely to race with static destructors for variables it accesses, typically through other libraries. This is a particularly subtle point, since older thread libraries were generally designed as C libraries, and the issue does not arise there.

Since older C++ standards do not support `quick_exit`, there appears to be no safe way to let a thread run until process exit, without shutting it down explicitly. Thus, a call like `pthread_detach()` should be avoided with C++ code. Note that some older C++ thread libraries, notably Boost [16], will by default detach a thread that still exists when the corresponding thread object is destroyed. This is extremely unsafe in the presence of exceptions and threads that return results in function-scope variables [5]. We believe C++ threads should *always* be joined before thread objects are destroyed.

Older C++ dialects generally do not support convenient destruction of `thread_local` objects. This avoids a number of issues related to races introduced by those destructors.

References

1. S.V. Adve. Designing memory consistency models for shared-memory multi-processors. PhD thesis, University of Wisconsin-Madison, Madison, WI, 1993.

2. S. Adve and H.-J. Boehm. Memory models: A case for rethinking parallel languages and hardware. *Communications of the ACM*, 53, 8, pp. 90–101, August 2010.

3. P. An, A. Jula, S. Rus, S. Saunders, T. Smith, G. Tanase, N. Thomas, N. Amato, and L. Rauchwerger. STAPL: An adaptive generic parallel C++ library. In *Workshop on Languages and Compilers for Parallel Computation (LCPC)*, Kumberland Falls, KY, August 2001.

4. H.-J. Boehm. Threads cannot be implemented as a library. In *Proceedings of the 2005 ACM SIGPLAN Conference on Programming Language Design and Implementation (PLDI)*, Chicago, IL, pp. 261–268, 2005.

5. H.-J. Boehm. N2802: A plea to reconsider detach-on-destruction for thread objects. http://www.open-std.org/jtc1/sc22/wg21/docs/papers/2008/n2802.html, 2008.

6. H.-J. Boehm and S. Adve. Foundations of the C++ concurrency memory model. In *Proceedings of the Conference on Programming Language Design and Implementation*, Tucson, AZ, pp. 68–78, 2008.

7. CilkArts. Cilk++ solution overview. http://software.intel.com/en-us/articles/intel-cilk-plus/ (Accessed August 4, 2011).

8. M. Frigo, C.E. Leiserson, and K.H. Randall. The implementation of the Cilk-5 multithreaded language. In *Proceedings of the ACM SIGPLAN' 98 Conference on Programming Language Design and Implementation (PLDI)*, Montreal, Quebec, Canada, pp. 212–223, 1998.

9. IEEE and The Open Group, POSIX.1-2008 (The Open Group Base Specifications Issue 7 and IEEE Std 1003.1-2008) IEEE Standard 1003.1-2001. IEEE, 2001.

10. Intel. *Intel(R) Threading Building Blocks: Reference Manual*. Intel, Document number 315415-001US, 2009.

11. James Reinders. *Intel(R) Threading Building Blocks: Outfitting C++ for Multicore Parallelism, by James Reinders*. O'Reilly, Sebastopol, CA, 2007.

12. L. Lamport. How to make a multiprocessor computer that correctly executes multiprocess programs. *IEEE Transactions on Computers*, C-28(9):690–691, 1979.

13. J. Manson, W. Pugh, and S. Adve. The Java memory model. In *Proceedings of the 32nd ACM Symposium on Principles of Programming Languages*, Long Beach, CA, 2005.

14. P. McKenney and R. Silvera. Example power implementation for C/C++ memory model. C++ standards committee paper WG21/N2745 = 08-0255, http://www.open-std.org/JTC1/SC22/WG21/docs/papers/2008/n2745.html, August 2008.

15. J.G. Mitchell, W. Maybury, and R. Sweet. Mesa language manual. Technical Report CSL-79-3, Xerox Palo Alto Research Center, Palo Alto, CA, 1979.

16. A. Williams. What's new in Boost Threads? *Dr. Dobbs Journal*, October 2008.

Chapter 5

Parallelism in .NET and Java

Judith Bishop

Contents

5.1 Introduction

.NET and Java are both platforms supporting object-oriented languages, designed for embedded systems and the Internet. As such they include libraries for the concurrent execution of separate threads of control, and for communication between computers via typed channels and various protocols. Just using any of these libraries in either language will not, however, automatically cause a program to speed up on a multicore machine. Moreover before it requires very careful programming to avoid making the classic errors when trying to maintain mutual exclusion of shared variables, or to achieve correct synchronization of threads without running into deadlock. In this chapter, we describe and discuss the advances that have been made in the past few years on both the .NET and Java platforms in making concurrent and parallel programming easier for the programmer, and ultimately more efficient.

Concurrent and parallel are terms that are often, incorrectly, used interchangeably. Since this book is about parallelism, we need to define it and explain why it is different to concurrency. *Concurrency* occurs naturally in all programs that need to interact with their environment: multiple threads handle different aspects of the program and one can wait for input while another is computing. This behavior will occur even on a single core. Successful concurrency requires fair scheduling between threads.

Parallelism is defined differently: the threads are intended to execute together and the primary goal is speedup. The wish is that by adding *n* cores, the program will run *n* times faster. Sometimes this goal can almost be achieved, but there are spectacular failures as well [1], and many books and papers spend time on how to avoid them [2–4].

Parallelism uses unfair scheduling to ensure all cores are loaded as much as possible. This chapter describes new libraries incorporated into familiar languages that help to avoid these pitfalls by representing common patterns at a reasonably high level of abstraction. The value of using these libraries is that *if one has a program that needs speeding up*, the programming is easier, the resulting code is shorter, and there is less chance of making mistakes. On the other hand, applying these techniques to tasks that are small, the overhead will most probably make it run slower.

5.1.1   Types of Parallelism

Both .NET and Java provide, in different ways, for different types of parallelism, that is

- *Data parallelism*: Used when the same operation has to be done on all elements of a collection, or for a fixed number of iterations. The libraries supply methods that mimic `for` and `foreach` for this purpose.

- *Task parallelism*: Used when entirely different operations are performed on either the same or different sets of data. The program can *fork* a new task for each separate operation, and they eventually all *join* up when they have completed.

- *Embarrassingly parallel*: If the iterations do not access shared data in memory or files, then the threads that implement them can be very efficiently implemented on the underlying framework. Many scientific problems have solutions with this property.

- *Granularity: Fine-grained* parallelism has many small tasks; *coarse-grained* parallelism has few large tasks. The granularity of a program affects choices that are made for parallelizing the program, and for the speedup on a given computer. For example, having only three large tasks on a 128 core machine could well only achieve a speedup of 3. Having 100 small tasks on a four core computer is a typical challenge faced by programmers as they strive for a speedup of 4 by switching between the tasks efficiently.

5.1.2 Overview of the Chapter

The chapter starts with a discussion of the two offerings available on the .NET 4 platform—TPL (Task Parallel Library) and PLINQ (Parallel Language-INtegrated Query). Then we look at Java's concurrency package available now in Java 5 and the proposals for Java 7, namely, jsr166z (ForkJoin) and extra166z (ParallelArray). In each case, we present examples and consider a variety of performance results from the literature.

5.2 .NET Parallel Landscape

Within the Common Language Runtime (CLR) team at Microsoft and Microsoft Research, there was considerable activity from 2007 onward to build on the .NET platform a set of libraries to support parallelism, released after much testing with Visual Studio 2010 in April 2010. The main guide to the libraries is at [5]. A discussion on the theory on which the libraries are based was presented at OOPSLA in 2009 [6], an extensive white paper on how to use them is at [4] and there are two books devoted to them [3,7].

We talk in this chapter of .NET because the libraries are defined at that level (Figure 5.1). That means that they are available to languages that run with managed code, in particular C#, F#, and Visual Basic. In the documentation available for the libraries, examples are given in all these languages. In this chapter, we shall concentrate on C#, as being closest to Java. C++ has a different set of libraries [8] because C++ does not run in managed code.

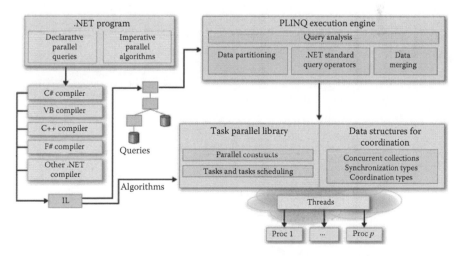

FIGURE 5.1: The .NET Framework parallel extensions. (From MSDN, Parallel programming with .NET, http://msdn.microsoft.com/en-us/library/dd460693% 28VS.100%29.aspx, 2010. With permission.)

The two libraries that emerged in the 2007–2010 timeframe for all .NET managed languages are as follows:

1. *TPL*. TPL enables programmers to easily express *task parallelism* in their programs, without equating tasks with threads in C# or with cores on their computer. The library is specifically intended for speeding up processor-bound computations. It is also intended to be a general abstraction for representing arbitrary asynchronous operations, whether compute-bound or I/O-bound. In the future, it is likely to become the standard .NET representation for asynchronous operations on multicore architectures [4].

2. *PLINQ*. PLINQ extends normal LINQ (Language INtegrated Query) syntax, which is similar to SQL, with additional operators for parallel programming for objects. PLINQ is specifically aimed at exposing *data parallelism* by use of queries, thereby leading to high-performance computing on the desktop by using all available cores efficiently [9,10].

5.3 Task Parallel Library

The Task Parallel Library (TPL) is part of the .NET Framework 4 and relies on a new task scheduler that is integrated with the .NET ThreadPool [11,12]. The TPL scales the degree of parallelism dynamically to most efficiently use all the cores that are available. In addition, the TPL handles the partitioning of the work, the scheduling of threads on the ThreadPool, cancellation support, state management, and other low-level details.

The TPL aims to make it easier to write managed code in C# (as well as other .NET languages) that can automatically harness the power of multicore processors. It provides methods to perform actions in parallel, and to handle automatic load balancing of the actions onto available cores. Using TPL, the programmer does not have to know how many cores there are, nor how the program is partitioned onto them, nor how any communication between the underlying threads is achieved.

Many attempts at parallel libraries have faltered because they require changes to the language (and hence the compiler) or they use complex comment or pragma syntax (e.g., OpenMP [13]) or they simply use libraries with hard-to-remember parameter lists (e.g., MPI [14]). Using a more innovative method calling syntax, but still without requiring language extensions, the researchers behind the .NET team managed to create a much more palliative and user-friendly approach. The key to the syntax is a heavy use of delegates, or ultimately, lambda expressions.

5.3.1 Basic Methods—For, ForEach, and Invoke

For simple iterative, data parallelism, TPL's approach is to provide parallel For and ForEach methods that indicate that the set of statements should be spread over multiple cores, but do not say how many:

```
Parallel.For (0, 100, delegate(i) {
    // do some significant work, perhaps using i
});
```

For is a static method of the System.Threading.Tasks.Parallel class. It takes three arguments, the last of which is a delegate expression representing the task to execute. i is a free variable that is passed into the body of the delegate at each iteration. The body of the delegate extends over the braces and becomes the task of the whole loop, which can be replicated and parallelized.

This syntax is new to ordinary programmers, but succinct and powerful. The creation of delegates can also be expressed as a lambda expression using the new => operator:

```
Parallel.For (0, 100, i => {
    // do some significant work, perhaps using i
});
```

If i is not going to be used, then the following overloaded version of For is appropriate:

```
Parallel.For (0, 100, () => {
    // do some significant work
});
```

The Parallel class has other methods: ForEach and Invoke. In keeping with the object-oriented nature of .NET languages, ForEach can be used to iterate over any enumerable collection of objects, which could be an array, list etc. The basic structure of the parallel version, as compared to the sequential one, is as follows:

```
// Sequential
foreach (var item in sourceCollection) {
Process(item);
}

//Parallel
Parallel.ForEach (sourceCollection, item =>
Process(item)
);
```

There are also overloads of ForEach that provide the iteration index to the loop body. However, the index should not be used to perform access to elements via calculated indices, because then the independence of the operations, potentially running on different cores, would be violated.

Invoke is the counterpart of the Par construct found in earlier languages [15,16]. This method can be used to execute a set of operations, potentially in parallel. As with the bodies of the For and ForEach, no guarantees are made about the order in which the operations execute or whether they do in fact execute in parallel. This method does not return until each of the provided operations has completed, regardless of whether completion occurs due to normal or exceptional termination. Invoke is illustrated in this example of a generic parallel Quicksort, programmed recursively, with recourse to Insertion sort when the lists are small:

```
// Example of TPL Invoke
// From [6]

static void ParQuickSort<T>(T[] domain, int lo, int hi)
                where T: IComparable<T> {
  if (hi - lo <= Threshold) InsertionSort(domain, lo, hi);
  int pivot = Partition(domain, lo, hi);
  Parallel.Invoke(
    delegate {ParQuickSort(domain, lo, pivot - 1);},
    delegate {ParQuickSort(domain, pivot + 1, hi);}
  );
}
```

Since Quicksort is recursive, a lot of parallelism is exposed because every invocation introduces more parallel tasks, which will be placed on cores as available. Of course, if there are only two cores, most of the tasks will be executed sequentially. Although this example uses delegates, invoked actions can also be lambda expressions or defined static methods.

5.3.2 Breaking Out of a Loop

Although breaking out of loops is common in sequential programming, it is less so in parallel programming. Nevertheless, TPL provides two options for breaking out—Break and Stop—based on a ParallelLoopState object that can be provided as a loop argument.

Suppose we want to stop processing a collection of student records as soon as the grade of any of them is detected to be 100. Consider the following code:

```
ParallelLoopResult loopResult =
Parallel.ForEach(Students,)
(student, ParallelLoopState loop) => {
// Lots of processing per student here, then
if (student.grade == 100 {
    loop.Stop();
    return;
}
// Checking if any other iteration has stopped
if (loop.IsStopped) return;
});

Console.WriteLine("Someone achieved 100%"+
loopResult.IsCompleted);
```

In this example, finding out details of the top student would be difficult, since the iterations would be competing to write them to a shared variable declared outside the loop. The break method, on the other hand, will complete all the iterations below the one where the break occurs, and it can return its iteration number. Moreover, if there is more than one student with 100%, the earlier snippet will likely give different answers on different runs—such is non-determinism.

5.3.3 Tasks and Futures

For, ForEach, and Invoke provide a simple way of expressing parallelism by using threads provided by the underlying scheduler (the thread pool) and distributing them onto cores without the programmer expending any effort. However, there are cases when more intervention is required. In parallel parlance, a future is a stand-in for a computational result that is initially unknown, but becomes available later. The process of calculating the result occurs in parallel with other computations. Essentially, a future in .NET is a task that returns a value.

A variation of a future is a *continuation task* that automatically starts when other tasks, known as its antecedents, complete. In many cases, the antecedents consist of futures whose result values are used as input by the continuation task. An antecedent may have more than one continuation task.

The concept of futures is provided for in TPL by a class called System. Threading.Tasks.Task <TResult> assisted by the TaskFactory class that encodes some common Task patterns into methods that pick up default settings, which are configurable through its constructors. Commonly used methods are as follows:

Task

```
Start, ContinueWith, RunSynchronously, Wait.
```

Factory

```
StartNew, ContinueWhenAll, ContinueWhenAny, FromAsynch
```

Consider the simple task graph in Figure 5.2 where one arm has the function F1 and the other arm has the functions F2 and F3 in sequence. The two arms join to provide input to F4 (adapted from [3]).

One example for creating futures for this graph would be as follows:

```
Task<int> futureB =
Task.Factory.StartNew<int>(() => F1(a));
int c = F2(a);
int d = F3(c);
int f = F4(futureB.Result, d);
return f;
```

The future created for F1(a) starts first and then on a multicore machine, F2 can start as well. F3 will start as soon as F2 completes and F4 as soon as F1 and F3 complete (in any order).

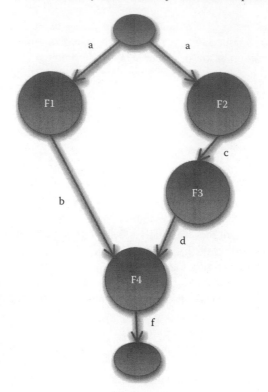

FIGURE 5.2: A task graph for futures. (From Campbell, C., et al., Parallel Programming with Microsoft® .NET: Design Patterns for Decomposition and Coordination on multicore Architectures, Microsoft Press, 167 pp. at http://parallelpatterns. codeplex.com/, 2010.

This example becomes more realistic when considering what happens to f. It is very common for one operation to invoke another and pass data to it and if the dependencies are made clear, then the runtime can do a better job of scheduling. This is the purpose of continuation tasks. Supposing f is being used to update a user interface, we would have the following:

```
TextBox myTextBox = ...;
var futureB = Task.Factory.StartNew <int>(() => F1(a));
var futureD = Task.Factory.StartNew <int>(() => F3(F2(a)));
var futureF = Task.Factory.ContinueWhenAll <int,int>(
                  new[] {futureB, futureD},
        (tasks) => F4(futureB.Result, futureD.Result));

futureF.ContinueWith((t) =>
  myTextBox.Dispatcher.Invoke(
        (Action)(() => {myTextBox.Text = t.Result.ToString();}))
    );
```

The task that invokes the action on the text box is now linked to the result of F4.

5.4 PLINQ

LINQ was introduced in the .NET Framework 3. It enables the querying of collections or data sources such as databases, files, or the Internet in a type-safe manner. LINQ to Objects is the name for LINQ queries that are run against in-memory collections such as `List <T>` and arrays and PLINQ is a parallel implementation of the LINQ pattern [10].

A PLINQ query in many ways resembles a nonparallel LINQ to Objects query. Like LINQ, PLINQs queries have deferred execution, which means they do not begin executing until the query is enumerated. The primary difference is that PLINQ attempts to make full use of all the processors on the system. It does this by partitioning the data source into segments, and then executing the query in each segment on separate worker threads in parallel on multiple processors. In many cases, parallel execution means that the query runs significantly faster.

Here is a simple example of selecting certain numbers that satisfy the criteria in an expensive method `Compute` and putting them in a collection of the same type as the source.

```
var source = Enumerable.Range(1, 10000);
// Opt-in to PLINQ with AsParallel
var evenNums = from num in source.AsParallel()
    where Compute(num) > 0
    select num;
```

The mention of `AsParallel` indicates that the runtime may partition the data source for computation by different cores. In PLINQ, the `var` declaration works out types at compile time.

PLINQ also has a `ForAll` method that will iterate through a collection in an unordered way (as opposed to using a normal `foreach` loop that would first wait for all iterations to complete). In this example, the results of the earlier query are added together into a bag of type `System.Collections.Concurrent.ConcurrentBag(Of Int)` that can safely accept concurrent bag operations.

```
evenNums.ForAll((x) => concurrentBag.Add(Compute(x)));
```

PLINQ will be of great benefit for graphics and data processing type applications of the embarrassingly parallel kind. It can of course also work in combination with the task synchronization provided by TPL. Consider the popular Ray Tracer example [17]. This method computes the intersections of a ray with the objects in a scene.

```
private IEnumerable<ISect> Intersections(Ray ray, Scene scene) {
    var things = from inter in obj.Intersect(ray)
        where inter != null
        orderby inter.Dist
        select inter;
    return things;
}
```

The example was originally written in C# 3.0 syntax and made references to the methods being called on each object more explicit:

```
// Example of LINQ
// Luke Hoban's Ray Tracer [17]
private IEnumerable <ISect> Intersections(Ray ray, Scene scene) {
    return scene.Things
        .Select(obj => obj.Intersect(ray))
        .Where(inter => inter != null)
        .OrderBy(inter => inter.Dist);
}
```

The `Intersections` method selects each object in the scene that intersects with the ray, forming a new stream of `Things`, then takes only those that actually intersect (the `Where` method), and finally uses the `OrderBy` method to sort the rest by the distance from the source of the ray. `Select`, `Where`, and `OrderBy` are methods that have delegates as parameters and the syntax used is the new => operator followed by the expression to be activated. It is the clever positioning of the dots that make this code so easy to read [18]. Query comprehension syntax was a natural advance in syntactic "sugar" for C# 4.0.

5.5   Evaluation

The TPL library contains sophisticated algorithms for dynamic work distribution and automatically adapts to the workload and particular machine [6]. An important point emphasized by Leijen et al. is that the primitives of the library express potential parallelism, but do not enforce it. Thus, the same code will run as well as it can on one, two, four, or more cores.

When a parallel loop runs, the TPL partitions the data source so that the loop can operate on multiple parts concurrently. Behind the scenes, the task scheduler partitions the task based on system resources and workload. When possible, the scheduler redistributes work among multiple threads and processors if the workload becomes unbalanced. Provision is also made for the user to redefine the task scheduler.

The TPL runtime is based on well-known work-stealing techniques and the scalability and behavior is similar to that of JCilk [19] or the Java fork-join framework [20]. In most systems, the liveness of a task is determined by its lexical scope, but it is not enforced. Tasks in TPL however are first-class objects syntactically and are allocated on the heap. A full discussion of the implementation, which includes the novel duplicating queues data structure, is given in [6].

In the presence of I/O blocking scheduling can also be an issue. By default, the TPL uses a hill-climbing algorithm to improve the utilization of cores when the threads are blocked. Customized schedulers can also be written. An example would be one that uses an I/O completion port as the throttling mechanism, such that if one of the workers does block on I/O, other tasks will be run in its place in order to allow forward progress on other threads to make use of that lost core.

In a similar way, PLINQs queries scale in the degree of concurrency based on the capabilities of the host computer. Through parallel execution, PLINQ can achieve significant performance improvements over legacy code for certain kinds of queries, often just by adding the `AsParallel` query operation to the data source. However, as mentioned earlier parallelism can introduce its own complexities, and not all query operations run faster in PLINQ. In fact, parallelization can actually slow down certain queries if there is too much to aggregate at the end of the loop, or if a parallel PLINQ occupies all the cores and other threads cannot work. This latter problem can be overcome by limiting the degree of parallelism. In this example, the loop is limited to two cores, leaving any others for other threads that might also be running.

```
var source = Enumerable.Range(1, 10000);
// Opt-in to PLINQ with AsParallel
var evenNums = from num in source.AsParallel()
                   .WithDegreeOfParallelism(2)
    where Compute(num) > 0
    select num;
```

In documented internal studies, TPL holds its own against hand-tooled programs and shows almost linear speedup for the usual benchmarks [6]. We converted the

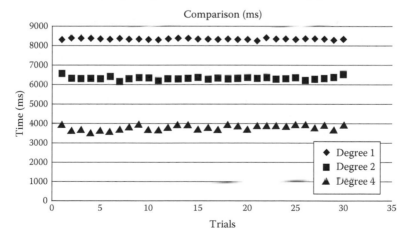

FIGURE 5.3: Speedup of the Ray Tracer.

Ray Tracer example found at [17] to TPL and PLINQ and had the speedups shown in Figure 5.3.

The machine was a 4x 6-core Intel Xeon CPU E7450 @ 2.4 GHz (total of 24 cores) with RAM: 96GB. The operating system was Windows Server 2008 SP2 64-bit (May 2009) running Visual Server 2010 with its C# compiler (April 2010).

5.6   Java Platform

Java was one of the first object-oriented language to include support for user-programmed (and hence lightweight) threads in 1997. Threads are available to run in parallel and, depending on the number of cores available and the number of competing threads, some of those threads actually *will* run in parallel on different cores, while in other cases the illusion of simultaneous execution will be achieved by having the operating system switching threads in and out of the available cores (even if it is only one). For parallel programming, however, the thread model is not an easy one to use, because it requires the programmer to handle all the synchronization and locking associated with shared variables. In 1999, Doug Lea wrote the seminal book on concurrency in Java [2] with a second edition in 1999, and his proposals for advanced thread usage [21], known as JSR166, were included in Java 5 in 2004. These are a set of concurrency utilities, including a thread pool, synchronizers, atomic variables, and other utilities, that are now included in the `java.util.concurrent` package.

Programming parallel solutions at this level is still quite a challenge, and Lea has for many years been working on a further lightweight task framework for Java known as ForkJoin (FJ and also JSR166y) [22]. Most of the framework is targeted for release with the Java 7 platform in July 2011. In the interim, many research projects blossomed to fill the void, such as PJ [23], MPJ Express [24], and JCilk [25].

5.6.1   Thread Basics

Java's thread model is based on `Runnable` objects. `Runnable` is an interface, with the single `run()` method that the user provides. So, for example, to display all numbers that satisfy a criteria and separately reassure the user that the program is still running, we might have the following:

```
Runnable reassure = new Runnable() {
    public void run() {
      try {
        while (true) {
          System.out.println(''Working ...'');
      } catch (InterruptedException iex) {}
    }
};
```

```
Runnable satisfiesCompute = new Runnable() {
   public void run() {
      try {
         for (int i; i<10000; i++) {
            if (Compute(i))
               System.out.println(i + ''is good'');
         }
      }
      catch (InterruptedException iex) {}
   }
};
```

Now, to run the preceding two tasks simultaneously, we create a `Thread` object for each `Runnable` `object`, then call the `start()` method on each `Thread`:

```
Thread r1 = new Thread(reassure);
Thread r2 = new Thread(r1);
r1.start();
r2.start();
```

A call to `start()` spawns a new thread that begins executing the task that was assigned to the `Thread` object at some time in the near future. Meanwhile, control returns to the caller of `start()`, and the second thread starts. So there will be at least *three* threads running in parallel: the two just started, plus the main. (In reality, the JVM will tend to have extra threads running for housekeeping tasks such as garbage collection, although they are essentially outside of the program's control.)

There are several other ways of programming the earlier solution, and together with standard synchronization primitives such as `wait` and `notify`, threads are adequate for many concurrency applications. However, the model is of a too low level for expressing solutions to well-known data parallelism or task parallelism problems that need to harness the power of multicores for involve arrays processing involving even simple synchronization.

5.6.2 `java.util.concurrent`

The `java.util.concurrent` package in Java 5 [26] provides classes and interfaces aiming at simplifying the development of concurrent and parallel applications by providing high-quality implementations of their common building blocks. The package includes classes optimized for concurrent access, including the following:

1. *Task scheduling framework*: The `Executor` framework is for standardizing invocation, scheduling, execution, and control of asynchronous tasks according to a set of execution policies, including queue length limits and saturation policy to improve the stability of applications by preventing runaway source consumption.

2. *Concurrent collections*: Provided with high-quality concurrent implementations for Map, List, and Queue and a new interface for BlockingQueue.

3. *Atomic variables*: For automatically manipulating single variables, which can be primitive types or references.

4. *Synchronizers*: General-purpose synchronization classes that facilitate coordination between threads that have been added including semaphores, mutexes, barriers, latches, exchangers and locks

5. *Nanosecond-granularity timing*: A System.nanoTime method that enables access to a nanosecond-granularity time source for making relative time measurements.

Based on [26], we can summarize these various elements. An Executor is a simple standardized interface for defining custom thread-like subsystems, including thread pools, asynchronous I/O, and lightweight task frameworks. Depending on which concrete Executor class is being used, tasks may execute in a newly created thread, an existing task-execution thread, or the thread calling execute(), and may execute sequentially or concurrently. For example, the following code will work through a pool of numWorkers threads to process a queue of patients:

```
logger.info("Starting benchmark");
threadExecutor = Executors.newFixedThreadPool(numWorkers);
long elapsedTimeMillis = System.currentTimeMillis();
while (!patient.isEmpty()) {
  threadExecutor.execute(pat.poll());
}
// Wait for termination
threadExecutor.shutdown();
```

Next up, the ExecutorService provides a more complete asynchronous task-execution framework. An ExecutorService manages queuing and scheduling of tasks, and allows controlled shutdown. The threadExecutor presented earlier, defined during the warm-up to this code, is declared as one of the following:

```
ExecutorService threadExecutor =
        Executors.newFixedThreadPool(numWorkers);
```

The ScheduledExecutorService subinterface adds support for delayed and periodic task execution. ExecutorServices provide methods arranging asynchronous execution of any function expressed as Callable, the result-bearing analog of Runnable. ExecutorCompletionService assists in coordinating the processing of groups of asynchronous tasks.

There are also thread pool versions of the executors, and factory methods for the most common kinds of configurations, as well as a few utility methods for using them. The concrete class FutureTask provides a common extensible implementation of `Future`. It returns the results of a function, allows determination of whether execution has completed, and provides a means to cancel execution.

The `ConcurrentLinkedQueue` class supplies an efficient scalable thread-safe non-blocking FIFO queue. Five implementations in `java.util.concurrent` support the extended `BlockingQueue` interface that defines blocking versions of `put` and `take`: `LinkedBlockingQueue`, `ArrayBlockingQueue`, `SynchronousQueue`, `PriorityBlockingQueue`, and `DelayQueue`. The different classes cover the most common usage contexts for producer-consumer, messaging, parallel tasking, and related concurrent designs.

5.7 Fork-Join Framework

Java 7 includes a new framework for more fine-grained parallelism [26] in the package `java.util.concurrent.forkjoin`. The package was developed by the JSR 166 Expert Group and can also be used with Java 6. The primary coordination mechanism is `fork()`, which arranges asynchronous execution, and `join()`, which does not proceed until the task's result has been computed. `invokeAll` (available in multiple versions) performs the most common form of parallel invocation: forking a set of tasks and waiting for them all to complete.

A `ForkJoinExecutor` is like an `Executor` in that it is designed for running tasks, except that it is specifically designed for computationally intensive tasks that do not ever block except to wait for another task being processed by the same `ForkJoinExecutor`.

The fork-join framework supports several styles of `ForkJoinTasks`, including those that require explicit completions and those executing cyclically. The `RecursiveAction` class directly supports the style of parallel recursive decomposition for non-result-bearing tasks; the `RecursiveTask` class addresses the same problem for result-bearing tasks. Finally, there is the `Phaser` class for a reusable synchronization barrier, similar in functionality to `CyclicBarrier` and `CountDownLatch` but supporting more flexible usage. Unlike the case for other barriers, the number of parties registered to synchronize on a phaser may vary over time [26].

Consider the example from [20] that illustrates some of these classes and methods:

```
public class MaxWithFJ extends RecursiveAction {
    private final int threshold;
    private final SelectMaxProblem problem;
    public int result;
    public MaxWithFJ(SelectMaxProblem problem,
        int threshold) {
        this.problem = problem;
        this.threshold = threshold;
    }
```

```
protected void compute() {
  if(problem.size < threshold)
    result = problem.solveSequentially();
  else {
      int midpoint = problem.size/2;
      MaxWithFJ left = new MaxWithFJ
              (problem.subproblem(0, midpoint), threshold);
      MaxWithFJ right = new MaxWithFJ(
              problem.subproblem(midpoint + 1, problem.size),
              threshold);
    invokeAll(left, right);
    result = Math.max(left.result, right.result);
  }
}

public static void main(String[] args) {
  SelectMaxProblem problem = ...
  int threshold = ...
  int nThreads = ...
  MaxWithFJ mfj = new MaxWithFJ(problem, threshold);
  ForkJoinExecutor fjPool = new ForkJoinPool(nThreads);
  fjPool.invoke(mfj);
  int result = mfj.result;
  }
}
```

The `RecursiveAction` class is extended to make more specialized tasks and it
has one method, `compute`, that is intended to be overridden. The `main` method sets
up a special thread pool of the class `ForkJoinExecutor`, which is specifically
designed for computationally intensive tasks that do not ever block except to wait
for another task being processed by the same `ForkJoinExecutor`. The next line
commences performing the first `mfj` task by calling its compute method. Unlike C#,
Java does not have delegates or lambdas, so the association between the thread call
and the tasks' execution cycle has to be specifically by name. The task awaits its
completion, returning its result, or throws an (unchecked) exception if the underlying
computation did so.

5.7.1 Performance

Goetz quotes the following results for the performance of the preceding program
[20]. Table 5.1 shows some results of selecting the maximal element of a 500,000
element array on various systems and varying the threshold at which the sequential
version is preferred to the parallel version. For most runs, the number of threads in
the fork-join pool was equal to the number of hardware threads (cores times threads-
per-core) available. The numbers are presented as a speedup relative to the sequential
version on that system.

TABLE 5.1:   Results of running selectMax on 500k element arrays on various systems.

	Threshold = 500k	Threshold = 50k	Threshold = 5k	Threshold = 500	Threshold = 50
Pentium-4 HT (2 threads)	1.0	1.07	1.02	.82	.2
Dual-Xeon HT (4 threads)	.88	3.02	3.2	2.22	.43
8-way Opteron (8 threads)	1.0	5.29	5.73	4.53	2.03
8-core Niagara (32 threads)	.98	10.46	17.21	15.34	

Source: Goetz, B., Java theory and practice: Stick a fork in it, http://www.ibm.com/developerworks/java/library/j-jtp11137.html, November 2007.

The results show that on two cores, the program actually slows down badly when it is forced to perform mainly in parallel (the last column with threshold 50). For four cores, the speedup is about three when the threshold is chosen well.

Lea [22] explains that ForkJoin pays off if

- There are lots of elements with small or cheap operations

- Each of the operations is time-consuming and there are not a lot of elements

In the tests on the right-hand columns, the operations were not sufficiently time-consuming. However, the principal benefit of using the fork-join technique is that it affords a portable means of coding algorithms for parallel execution. The programmer does not have to be aware of how many cores will be available and the runtime can do a good job of balancing work across available workers, yielding reasonable results across the wide range of hardware that Java runs on [20].

5.8   **ParallelArray** Package

A package that was proposed with ForkJoin, for sorting and searching and selection problems, is the ParallelArray. ParallelArray objects represents a collection of structurally similar data items, and there are methods similar to PLINQ to specify data operations in SQL and hide the mechanics of how the operations are implemented. The aggregate processing must have a regular structure, expressed in terms of apply, reduce, cumulate, sort, and so on to match the methods available [27]. As an example, consider the following:

```
// Example in Java of ForkJoin ParallelArray-
// Accessing a student database
// Adapted and shortened from [28]
```

```
import jsr166y.forkjoin.*;

public static void main(String[] args) {
    int classSize = new Integer(args[0]);
    Student[] currentClass =  generateStudents(classSize);
    ForkJoinExecutor fjPool = new  ForkJoinPool(25);
    ParallelArray<Student> processor =
      ParallelArray.createUsingHandoff(currentClass, fjPool);

  System.out.println("Sorted by student ID:");
    processor
      .sort(studentIDComparator)
      .apply(printOutStudent);
  System.out.println("");

  System.out.println("Graduates this year:");
    processor
        .withFilter(isSenior)
        .withFilter(hasAcceptableGPA)
        .apply(printOutStudent);
  System.out.println("");
  }
```

It is very important to note that the parameters to the `sort`, `apply`, and `withFilter` methods are not methods themselves: they are objects and the action happens because there is a method of an agreed name that is overridden. In this case it is `op`. So, for example, the class for `printOutStudent` is as follows:

```
private static final Ops.Procedure<Student> printOutStudent =
    new Ops.Procedure<Student>() {
      public void op(Student s) {
        System.out.println("\t" + s);
      }
    };
```

`ParallelArray` has several limitations. For example, filter operations must be specified before mapping operations. However, but its main purpose is to free developers from thinking about how the work can be parallelized. Once a program is written, reasonable performance can be expected. A speedup of 3x on four cores could be obtained for the preceding program without any tuning.

5.9 Conclusion

Although in the common mind the .NET and Java platforms seem to be very similar, the way they have been able to develop has been completely different, and in certain critical cases, such as support for parallel processing (or in the past, generics), this difference has led to one getting ahead of the other.

Java has a community-based development model. Researchers from around the world propose alternate solutions to problems. The advantage is that once they get to the stage of adoption by Sun (now Oracle), there is a ready community of users. .NET is supported more by its own researchers who have a faster track to getting new ideas adopted. The downside is that a user community is not automatically built at the same time. The effect of these business factors on technology can be long-lasting.

In making use of concurrency, we still need to take into consideration Amdahl's law, which states that the amount of speedup for any code is limited by the amount of code that can be run in parallel. A paper that strikes a warning is that of Hill and Marty [29], where they explain that there are at least three different types of multicore chip architectures, which they call symmetric, asymmetric, and dynamic, and that the performance across multiple cores for standard functions was quite different in each case. It will be a challenge worth pursuing to see how the language features described in the preceding text stand up to the new machines, not just the test cases on the desktop.

Acknowledgments

My thanks to Ezra Jivan, while in the Polelo Group, University of Pretoria, for his assistance with the programming; to Mauro Luigi Dragos of Politecnico di Milano for checking and retuning the programs; and to Stephen Toub and Daan Leijen, Microsoft, for their careful reading of the draft and their expert comments. Section 5.5 relies on the work done with Campbell et al. [3]. Work for this chapter was initially done when the author was at the University of Pretoria, South Africa, and supported by a National Research Foundation grant.

References

1. Holub, A., Warning! Threading in a multiprocessor world, JavaWorld, http://www.javaworld.com/jw-02-2001/jw-0209-toolbox.html, September 2001.

2. Lea, D., *Concurrent Programming in Java: Design Principles and Patterns*, 2nd edn. Addison Wesley, Reading, MA, 1999.

3. Campbell, C., R. Johnson, A. Miller, and S. Toub, *Parallel Programming with Microsoft® .NET: Design Patterns for Decomposition and Coordination on Multicore Architectures*, Microsoft Press, 167pp, at http://parallelpatterns.codeplex.com/, 2010.

4. Toub, S., Patterns of parallel programming—Understanding and applying patterns with the .NET framework 4 and Visual C#, White Paper, 118pp.,

http://www.microsoft.com/downloads/details.aspx?FamilyID=86b3d32b-ad26-4bb8-a3ae-c1637026c3ee&displaylang=en, 2010.

5. MSDN, Parallel programming with .NET, http://msdn.microsoft.com/en-us/library/dd460693%28VS.100%29.aspx, 2010.

6. Leijen, D., W. Schulte, and S. Burckhardt, The design of a task parallel library, *Proc. 24th ACM SIGPLAN Conference on Object Oriented Programming Systems Languages and Applications (OOPSLA '09)*, Orlando, FL, pp. 227–242. DOI = http://doi.acm.org/10.1145/1640089.1640106

7. Freeman, A., Pro *.NET 4 Parallel Programming in C# (Expert's Voice in .NET)*, APress, New York, 328pp. 2010.

8. Campbell, Colin, A. Miller, *Parallel Programming with Microsoft® Visual Studio C++®: Design Patterns for Decomposition and Coordination on Multicore Architectures*, Microsoft Press, 208pp, at http://parallelpatterns.codeplex.com/, 2011.

9. Duffy, J. and E. Essey, Running queries on multi-core processors. *MSDN Magazine*, http://msdn.microsoft.com/en-us/magazine/cc163329.aspx, October 2007.

10. MSDN, Parallel LINQ (PLINQ), http://msdn.microsoft.com/en-us/library/dd460688.aspx, 2010.

11. MSDN, Task Parallel Library overview, http://msdn.microsoft.com/en-us/library/dd460717(VS.100).aspx, July 2009.

12. Stephens, R., Getting started with the .NET Task Parallel Library, DevX.com, http://www.devx.com/dotnet/Article/39204/1763/, September 2008.

13. OpenMP, http://openmp.org/wp/, 2010.

14. Gregor, D. and A. Lumsdaine, Design and implementation of a high-performance MPI for C# and the common language infrastructure, *Proc. 13th ACM SIGPLAN Symposium on Principles and Practice of Parallel Programming*, Salt Lake City, UT, pp. 133–142, 2008.

15. *Handel-C Language Reference Manual*. Embedded Solutions Limited, 1998.

16. occam 2.1 Reference Manual. SGS-THOMSON Micro-electronics Limited, 1995.

17. Hoban, L., A ray tracer in C# 3.0, blogs.msdn.com/lukeh/archive/2007/04/03/a-ray-tracer-in-c-3-0.aspx

18. Bierman, G. M., E. Meijer, and W. Schulte, The essence of data access in Comega: The power in the dot. *ECOOP 2005*, Glasgow, U.K., pp. 287–311.

19. Lee, I.-T. A, The jcilk multithreaded language. Masters thesis, MIT, also at http://supertech.csail.mit.edu/jCilkImp.html, September 2005.

20. Goetz, B., Java theory and practice: Stick a fork in it, http://www.ibm. com/developerworks/java/library/j-jtp11137.html, November 2007.

21. Lea, D., The java.util.concurrent Synchronizer framework, *PODC Workshop on Concurrency and Synchronization in Java Programs, CSJP'04*, July 26, 2004, St John's, Newfoundland, CA http://gee.cs.oswego.edu/dl/papers/aqs.pdf

22. Lea, D., A java fork/join framework. Java Grande, pp. 36–43, 2000.

23. Kaminsky, A. *Building Parallel Programs: SMPs, Clusters, and Java.* Cengage Course Technology, Florence, KY, 2010. ISBN 1-4239-0198-3.

24. Shafi, A., B. Carpenter, and M. Baker, Nested parallelism for multi-core HPC systems using Java, *Journal of Parallel and Distributed Computing*, 69(6), 532–545, June 2009.

25. Danaher, J. S., I.-T. A. Lee, C. E. Leiserson, Programming with exceptions in JCilk, *Science of Computer Programming*, 63(2), 147–171, 2006.

26. http://download-llnw.oracle.com/javase/6/docs/api/java/util/concurrent/package-summary.html

27. http://www.javac.info/jsr166z/jsr166z/forkjoin/ParallelArray.html

28. Neward, T., Forking and joining Java to maximize multicore power, DevX.com, http://www.devx.com/SpecialReports/Article/40982, February 2009.

29. Hill, M. D. and M. R. Marty, Amdahl's law in the multicore era, http://www. cs.wisc.edu/multifacet/amdahl/ (for interactive website) and IEEE Computer, pp. 33–38, July 2008.

Chapter 6

OpenMP

Barbara Chapman and James LaGrone

Contents

6.1   Introduction

OpenMP (OpenMP 2009), the Open standard for shared memory MultiProcessing, was designed to facilitate the creation of programs that are able to exploit the features of parallel computers where memory is shared by the individual processor cores. Widely supported by mainstream commercial and several open source compilers, it is well suited to the task of creating or adapting application codes to execute on platforms with multiple cores (Intel 2009, Open64 2005, Portland 2009, Sun 2005). OpenMP is intended to facilitate the construction of portable parallel programs by enabling the application developer to specify the parallelism in a code at a high level, via an approach that moreover permits the incremental insertion of its constructs. However, it also permits a low-level programming style, where the programmer explicitly assigns work to individual threads.

In this chapter, we will first give an overview of the major features of OpenMP and its usage model, as well as remark upon the process by which it has been designed and is further evolving. In Section 6.2, we will then introduce the basic constructs of OpenMP and illustrate them using several short examples. In Section 6.3, we then describe the manner in which OpenMP is implemented. One of the greatest challenges facing any application developer is learning how to overcome performance problems. Some understanding of the strategy used to implement OpenMP, and a basic appreciation of the implications of a shared memory model, will help us to discuss the most important considerations in this regard; this is the topic of Section 6.4. We then briefly discuss some performance considerations before concluding with a recap of the main points.

6.1.1 Idea of OpenMP

OpenMP provides a particularly straightforward method of converting a sequential program written in C, C++, or Fortran for execution on a system with multiple cores, several processors, or both. In many cases, all the application developer must do is insert a few directives (or pragmas) into a preexisting source code and select the appropriate compiler option. The compiler will then "recognize" these directives and use the information they contain to translate the program for multithreaded execution. If the corresponding compiler option is not specified at translation time, then the directives will be ignored and a single-threaded, sequential code will be generated. Hence, if a few rules are observed, one and the same application code can be both sequential and parallel: this can be particularly useful for development and testing purposes. The nice thing about this approach is that the application developer does not have to explicitly create the code that will run on the different cores, or processors. This is the job of the OpenMP implementation. As a result, it can be one of the easiest, and fastest, means of converting existing applications for use on multicore platforms. Moreover, it is possible to convert parts of a program for multithreaded execution and to leave the rest of the code in sequential form. If almost all of the work of a program is in one loop, for instance, it might make sense to parallelize that loop using OpenMP's features and to leave the rest of the program untouched.

6.1.2 Overview of Features

The OpenMP Application Programming Interface (API) comprises a collection of directives for expressing the parallelism that is to be exploited in a computation, a small number of library routines that may be used to access or influence execution parameters, and some environment variables for setting default values (OpenMP 2008). It is prescriptive in the sense that the implementation will translate a program for multithreaded execution exactly according to the application developer's instructions. In particular, it will not attempt to extract additional parallelism from the code. Programs that rely primarily on directives for the specification of parallelism have a different flavor from those that mainly use the library routines to express the parallelism, and we will give an example of each later in the text. Typically, both directives and library routines are needed to create an OpenMP program.

The API provides a means for the programmer to explicitly create, assign work to, synchronize, and terminate a set of numbered, logical threads. The implementation will use this information to translate an OpenMP code into a collection of cooperating tasks that are assigned to the system-level threads on the target platform for execution. These statically defined tasks are transparent to the programmer; they will be executed in the order and manner that corresponds to the semantics of the OpenMP constructs used. However, the API also enables the application developer to explicitly specify tasks that will be dynamically created and scheduled for execution in an order that is not predetermined. The specification further provides a means to synchronize the actions of the various logical threads in order to ensure, in particular, that values created by one thread are available before another thread attempts to use them. It is also possible to ensure that only one thread accesses a specific memory location, or performs a set of operations, at a time without the possibility of interference by other threads. This is sometimes required in order to avoid data corruption. Last, but not least, the specification provides a number of so-called data attributes that may be used to state whether data is to be shared among the executing threads or is private to them. The logical threads may cooperate by reading and writing the same shared variables. In contrast, each thread will have its own local instance of a private variable, with its own value. This makes it relatively easy for the threads to work on distinct computations and facilitates efficiency of execution.

An OpenMP implementation typically consists of two components. The first of these is the compiler, created by extending a preexisting compiler that implements one or more of the base languages Fortran, C, or C++. It will translate the OpenMP constructs by suitably modifying the program code to generate tasks and manage the data. As part of this translation, it will insert calls to routines that will manage the system threads at run time and assign work to them in the manner specified by the application programmer. The custom runtime library that comprises these routines is the second component in an OpenMP implementation. Commercial compilers will use the most efficient means possible to start, stop, and synchronize the executing threads on the target platforms they support. Open source compilers will typically place a higher value on portability and may rely on Pthreads routines to accomplish this functionality.

Initially, an OpenMP program's execution begins with a single thread of control, just like a sequential program. When this initial thread encounters code that is to be executed in parallel, it creates (and becomes a member of) a team of threads to execute the parallel code; the original thread becomes the master of the team. The team of threads will join at the termination of the parallel code region, so that only the original master thread continues to carry out work. Hence OpenMP is directly based upon the fork-join model of execution. An OpenMP program may contain multiple code regions that are outside of any parallel constructs and tasks; collectively, they are known as the sequential part of the program and they will be executed by the initial thread. Since the members of a team of threads may encounter code that is to be executed in parallel, they may each create and join a new team of threads to execute it. This can lead to nested parallelism, where hierarchies of thread teams are created to perform computations (Jost et al. 2004, Chapman et al. 2006). As a result, several different teams of threads may be active at the same time. Most OpenMP constructs are applied to the *current team*, which is the particular team that is currently executing the region of code containing it.

6.1.3　Who Developed OpenMP? How Is It Evolving?

The OpenMP specification is maintained by the OpenMP Architecture Review Board (ARB), an international consortium of organizations that have a strong interest in furthering the use of this portable API, including most of the leading hardware vendors (OpenMP 2009). Most members also provide OpenMP implementations, but some are primarily users of this technology and an international group of researchers are also active contributors (Compunity 2009).

The original feature set, introduced late 1997 as version 1.0 of the OpenMP standard, was designed to work well with Fortran 77 application codes and was based upon a body of prior work that attempted to provide a portable shared memory programming interface for scientific and technical computing. AC/C++ binding was produced shortly thereafter, and both of them underwent minor extensions. Subsequently, the two distinct specifications were merged to produce version 2.5 of the standard. In 2008, features for specifying explicit tasks were introduced and the implementation model was described in terms of tasks (Ayguadé et al. 2009b). Other new features in this version 3.0 of OpenMP (OpenMP 2008) include more support for hierarchical parallelism and additional strategies for parallelizing loop nests.

The ARB is continuing its endeavors to enhance this API and to consider how to improve its support for multicore platforms (Curtis-Maury et al. 2005). Its members meet regularly to debate potential language extensions and to discuss related topics (Mattson 2003). The increasing thread count that can be supported by shared memory architectures, including multicore systems, the wider use of accelerators along with general-purpose CPUs, and the broad adoption of parallelism in our hardware— from laptops all the way to supercomputers—all require that the members of the ARB consider whether extensions to the current API may be required. There is a natural tension between the desire to maintain a relatively simple, high-level programming interface on the one hand, and the need to satisfy the expression of a broad variety of parallel computations, and to facilitate their exploitation on a variety of platforms, on the other hand. Thus, new features are chosen only after a careful evaluation of application developer needs and a weighing of alternatives. Nevertheless, we expect that some extensions will be made to the current specification as multicore technology evolves and is directly exploited by an increasing percentage of applications (Ayguadé et al. 2009a, Gan 2009, Larsen et al. 2009).

6.2 OpenMP 3.0 Specification

In the following sections, we introduce the most important features of the OpenMP 3.0 API and illustrate their usage. They may be used in conjunction with code written in one of the three base languages (Fortran, C, and C++). We do not attempt to provide a complete overview of these features, nor do we describe all of the rules that must be followed if they are to be used correctly. For full details of the language, the specification should be consulted.

We will begin our discussion by introducing the most common directives. In OpenMP, most directives are inserted immediately before a structured block of code, to which they will be applied. A block of code is structured if it has precisely one entry and one exit: as a consequence, the use of certain language features (such as a STOP statement in Fortran) is prohibited in conjunction with some directives. In Fortran, it is generally necessary to also explicitly mark the end of a structured block. Therefore, the Fortran binding for OpenMP includes end directives as well. A directive and the associated code are together known as a construct. Various clauses may be appended to a directive in order to convey additional information, or instructions, to the implementation. We describe the most important of these clauses in the following also.

6.2.1 Parallel Regions and Worksharing

The fundamental OpenMP directive is the `parallel` directive, which is used to mark a structured block of code that should be executed by multiple threads. The code that will be executed in parallel as a result of this directive is referred to as a parallel region. If the parallel region is subsequently executed on, say, four cores with two threads supported per core, then there will be eight instances of the code in the parallel region, all running at the same time (the actual number of threads used may be determined by the programmer). The implementation will accomplish this by creating eight tasks, each of which will be assigned to one of the system threads at runtime. But without any further adaptation, the tasks will be identical and thus each thread will perform exactly the same computations as all the others. Generally, we will want to assign some portion of the overall work to each of them, rather than replicating it. For this, the so-called worksharing directives are provided. Without proper coordination via worksharing or synchronization constructs, such replicated work may introduce data races.

```
/*****************************************************************
C code with a parallel region where each thread
reports its thread id.
*****************************************************************/
#pragma omp parallel
{
printf("Hello from thread %d.\n",omp_get_thread_num());
}
```

Possible output:

```
Hello from thread 3.
Hello from thread 6.
Hello from thread 2.
Hello from thread 7.
Hello from thread 0.
Hello from thread 5.
Hello from thread 1.
Hello from thread 4.
```

Example 6.1

This example shows each thread executing the same statement. Possible output with eight threads is shown.

The most frequently used worksharing directive is the `loop` directive. This directive distributes the iterations of the associated countable loop nest among the threads in the team executing the enclosing parallel region. Note that, to be countable, the number of iterations of the loop must be known at the start of its execution (in contrast to, say, a loop nest that traverses a linked list). The application programmer may optionally describe the method to be used for assigning loop iterations to threads. Another worksharing directive, the `sections` directive, enables the specification of one or more sequential sections of code that are each executed by a `single` thread. The single directive marks code that should be executed by precisely one thread. Finally, the `workshare` directive is provided for Fortran programmers to specify that the computations in one or more Fortran 90 array statements should be distributed among the executing threads.

```
!***********************************************************
! Fortran code with workshare and atomic constructs.
!***********************************************************
!$OMP PARALLEL
!$OMP WORKSHARE
          A = B - 2*C
          B = A/2
          C = B
!$OMP     ATOMIC
                R = R + SUM(A)
!$OMP END WORKSHARE
!$OMP END PARALLEL
```

Example 6.2

This workshare directive directs the parallel execution of each statement of array operations. The ATOMIC directive synchronizes writes to R to prevent a data race.

Each of these directives except the `workshare` directive may have one or more clauses appended to control its application. The applicable clauses differ depending on the kind of directive, but all of them allow for the specification of attributes for the data used in the construct. Additional clauses that may be added to the `parallel` directive can be used to state how many threads should be in the team that will execute it and to describe any conditions that must hold for the parallel region to be actually executed in parallel. If the specified conditions do not hold, the associated code will be executed by a single thread instead. Additional information that may be given via clauses in conjunction with the loop directive includes stating how many loops in the loop nest immediately following the directive will be parallelized, and specifying a loop schedule, which is a strategy for assigning loop iterations to threads.

One of the most powerful innovations in the OpenMP API is that it does not require the worksharing directives to be in the same procedure as the `parallel` directive. In other words, one or more procedure calls may occur within a `parallel` construct and any of the invoked procedures may contain OpenMP directives. Directives that are not in the same procedure as the enclosing `parallel` directive are known as orphaned directives. Note, too, that a procedure containing orphaned directives may also be invoked outside of any parallel region, i.e., in the sequential part of the code. In such a situation, the directive is simply ignored.

Since the threads in a team perform their work independently, there can be minor variations in the speed at which computations progress on the different threads. Thus, the relative order in which results are produced is in general not known a priori. Moreover, since the results of computations are often initially stored in memory that is local to the thread where they were computed, new values may not be immediately accessible to threads running elsewhere on the machine. In order to provide some guarantees on the availability of data that are independent of any hardware or operating system consistency mechanisms, both parallel regions and worksharing constructs are terminated by a barrier, at which point the threads wait until all of them have completed execution of the construct and results are made available to all threads. Subsequent computations may therefore safely assume their availability and exploit them. Since a barrier may introduce inefficiencies if some threads must wait for slower ones to complete their assigned portion of work, it is possible to override its insertion at the end of worksharing directives (but not at the end of parallel regions, where the barrier is mandatory). For this, the `nowait` clause is provided. However, if the barrier is omitted, there are no guarantees on the availability of new values of shared data. Note that it is part of the job of the application developer to ensure that there is no attempt to use data before it is known to be available. The implementation does not test for this kind of error in the use of OpenMP directives.

In addition to the barrier, there are a few other features in the API that trigger the updating of shared data so that all threads will have a consistent view of their values. In addition to writing back any new values to main memory, threads will then retrieve any new values for shared variables that they have copied into local memory. Such places in the code are *synchronization points*. Between synchronization points, code executing on different threads may temporarily have different values of shared data.

As a result, OpenMP has a relaxed consistency model (Adve and Gharachorloo 1996, Bronevetsky and de Supinski 2007).

6.2.1.1 Scheduling Parallel Loops

It is sometimes important to be able to influence the way in which iterations of a parallel loop are assigned to threads, since this may have a major impact on the performance of the loop (Ayguadé et al. 2003, Chandra et al. 2000, Chapman 2008). For this, a so-called schedule clause may be appended to the loop directive. It may specify a static, dynamic, guided, auto(matic), or runtime schedule. When a static schedule is applied, the iterations will be partitioned into a set of contiguous chunks that will be assigned to the threads in a round-robin fashion. If no chunk size is given, a roughly equal number of iterations is assigned to each thread in a single chunk. When the amount of work is about the same in each iteration of the parallel loop, then a static schedule will distribute work evenly among threads. However, if the amount of work per iteration varies widely, a dynamic or guided schedule might be more appropriate. Under a dynamic schedule, each thread will grab a chunk of iterations and return to grab another chunk when it has executed them. Note that in this case, if no chunk size is specified, the default is a single iteration. A guided schedule is like the dynamic one except that the size of the chunks varies: it is set to be a proportion of the iterations remaining to be performed and thus will continually decrease. The auto schedule gives the implementation complete freedom to choose a schedule, including applying schedules that cannot be described using OpenMP features. Finally, the runtime schedule will allow the actual schedule to be determined during execution using an environment variable, and a library routine is provided for this purpose (see Section 6.2.5). If the programmer does not supply a schedule clause, then an implementation-defined default schedule will be applied. Note that this is almost always the static schedule.

```
!**********************************************************
! Fortran code with parallel loop nest and collapse,
! schedule clauses.
!**********************************************************
!$omp parallel do collapse(2) schedule(static)
      do i=1,np
        do j=1,nd
          call random_number(x)
          pos(j,i) = box(j)*x
          vel(j,i) = 0.0
        acc(j,i) = 0.0
        enddo
      enddo
!$omp end parallel do
```

Example 6.3
Use of OpenMP directives with Fortran.

Our first code example shows how OpenMP can be used to parallelize a loop nest in a Fortran program. A parallel region has been identified via the insertion of two directives, one at the start marking its beginning and one at the end. Note that here the parallel directive and the loop directive, needed to specify that the following loops' iterations should be shared among the threads, have been combined in a single line. Since statements beginning with an exclamation mark are considered to be comments in Fortran 90, a compiler that does not implement OpenMP will consider these directives to be comments and will simply ignore them. This will also happen if the user does not invoke the compiler with any options required to instruct it to translate OpenMP. The directives are clearly recognizable as a result of the omp prefix which is used to mark them.

Here, the `parallel do` directive will result in the formation of a team of threads: the programmer has not specified how many threads to use, so that either a default value will be applied or a value that has been set prior to this will be utilized. Each thread in the team will execute a portion of the parallel loop that follows. Since the `collapse` clause has been specified with 2 as an argument, both of the following two loops will be collapsed into a single iteration space and then divided as directed by the `schedule` clause. Thus, the work in the iterations of the i and j loops, a total of $np*nd$ iterations of the loop body, will be distributed among the threads in the current team. The static schedule informs the compiler that each thread should receive one contiguous chunk of iterations. For example, if the current team has 4 threads and $np = nd = 8$, there are 64 iterations to be shared among the threads. Each of them will have 16 contiguous iterations. Thus, thread 0 might be assigned the set of all iterations for which i is 1 or 2. If a different mapping of iterations to threads is desired, a chunk size can be given along with the kind of schedule. With a chunk size of 8, our thread 0 will instead execute two contiguous sets of iterations. These might be all iterations for which i is either 1 or 5.

```
/************************************************************
    Basic matrix multiply using OpenMP.
 ************************************************************/
#pragma omp parallel private(i,j,k)
{
    #pragma omp for schedule(static)
        for (i = 0; i < N; i++)
         for (k = 0; k < K; k++)
          for (j = 0; j < M; j++)
            C[i][j] = C[i][j] + A[i][k]*B[k][j];
} /* end omp parallel */
```

Example 6.4
This shows a naïve matrix multiply. Arrays A, B, and C are shared by default. Loop counters are kept private.

6.2.2   Data Environment

By default, most data in an OpenMP program is shared among all of the executing threads. If shared data is modified by a thread, then the new value is guaranteed to be available to other threads immediately after the next barrier or other synchronization point in a program. However, as already mentioned, threads may also have private, or local data. Each thread has its own copy of each private object, and their values may differ; although the use of private data may increase the memory requirements of a code, access to it is generally highly efficient. Moreover, the availability of local copies of some data may sometimes reduce the amount of synchronization needed between multiple threads. Threads may also have static private data, known as *threadprivate* data, so long as certain restrictions hold (mainly, there must be the same number of threads in the executing thread teams and there must be no nested parallelism). The values of threadprivate objects are then preserved between different parallel regions.

Data declared to be *firstprivate* is a private variable that is moreover initialized with the value of the corresponding variable immediately prior to the construct where this attribute is specified. If this clause occurs together with a parallel directive, then the value of the corresponding variable on the master thread is applied. Variables may also be declared to be *lastprivate* or to be *reduction* variables. The former of these causes the last value of the corresponding private variable to be saved and assigned to the value in the enclosing construct. Note that in this context "last" means the value assigned in the last iteration that would be executed if the associated loops were performed sequentially, or the value assigned in the last piece of source code in a set of sections. Also, there may sometimes be overheads involved in determining which value is the last one.

Reduction operations, in which a number of values are combined, for example, to form their sum or to find the maximum value, are common in many algorithms. Without special handling, a loop containing a reduction would be sequential, since each iteration would form a result depending on that of the previous iteration. OpenMP allows these loops to be parallelized so long as the application developer specifies that the loop contains a reduction and indicates the variable and kind of reduction via the corresponding clause. The reduction options include dot products, Boolean operations, and bit operations. Note that user-defined reductions are not permitted.

Data-sharing attributes may be specified in the form of clauses that can be used in conjunction with a `parallel` directive, with worksharing directives (except workshare), and with a `task` directive (see Section 6.2.3). In the case of a reduction, the kind of reduction must also be given. For those variables that are not specifically assigned an attribute, defaults are applied. Rules governing defaults are designed to be intuitive. For example, the local data of procedures that are invoked inside parallel regions are generally private; formal parameters inherit the data-sharing attributes of the corresponding actual arguments. The iteration variable of a parallel loop is an important example of a variable that is private by default. The API allows the programmer to override default data attributes in most cases.

6.2.2.1 Using Data Attributes

Our next example illustrates the use of OpenMP in conjunction with C, which does not require the use of end directives and where instead of structured comments, OpenMP directives are expressed as pragmas. In this code, the application developer has inserted a parallel region directive in the main procedure, which we do not show. The threads in the team created to execute it will all call the procedure foo. The automatic variables p and i are private by default, so that each thread has its own private instance of them. The formal parameters m and n will have the same data sharing attributes as the corresponding actual arguments. When the parallel sections construct is encountered, each individual section will be executed by one of the threads. The use of the nowait clause means that there will be no barrier at the end of the construct. This allows each thread to proceed past the sections construct after executing its section without waiting for the other thread to complete its share. If there are no dependences in the code that require the threads to wait for each other, this is a suitable way to ensure that each thread may continue to do useful work without unnecessarily impeding performance.

```
/***********************************************************
   Orphaned OpenMP worksharing directive (sections) with
   a nowait clause.
 ***********************************************************/
int g;
void foo(int m, int n) {
int p, i;
#pragma omp sections firstprivate(g) nowait
{
    #pragma omp section
      {
       p = foo(g);
        for (i = 0; i < m; i++)
          do_stuff;
 }
#pragma omp section
{
 p = bar(g);
    for (i = 0; i < n; i++)
      do_other_stuff;
}
}
return;
}
```

Example 6.5
Use of OpenMP directives with C.

6.2.3 Explicit Tasks

The `task` directive provides another powerful and flexible means of describing the parallelism in a broad variety of applications. It is particularly suited to expressing the parallelism in algorithms where the amount of parallelism is not known in advance, for instance, when recursion is involved or when each element in a list or other pointer-based structure must be processed in some fashion. When this directive is encountered at run time, the structured block of code that it is associated with will be made ready for execution. This requires, in particular, that any current values of data that it uses during computation are retrieved. The task may then be immediately executed by an arbitrary thread, or it might have to wait until a thread is ready to do so. There are no guarantees on the order in which tasks are executed if they are created between the same pair of synchronization points. Therefore, the computation must not rely on any such ordering nor make any assumptions with respect to which thread a task will be assigned to.

Each individual task will be executed sequentially by just one thread. However, if another task construct is encountered during its execution, then it will be responsible for creating a new task that will be separately scheduled for execution. Since it is possible that very many tasks are created around the same time, the implementation is allowed to suspend the execution of a given task in order to perform other work for a while. By default, execution of a suspended task is subsequently resumed on the same thread. The reason for this is that a task might access data that is private to that thread. If this is not the case, then the programmer may specify that a task is untied. This means that, if suspended, the untied task may be resumed on any available thread. This might somewhat improve program performance.

One reason that this construct is so flexible is that it may appear anywhere in a parallel region, including within another task. In other words, tasks may be nested. They are therefore highly suitable for expressing the parallelism in a recursive algorithm. However, worksharing directives may not appear within a task unless a new parallel region has been created. An additional important restriction is that a task may not contain a barrier. (Given the arbitrary execution ordering and location, this would otherwise most likely lead to a deadlock.)

6.2.3.1 Using Explicit Tasks

The following code snippet illustrates a typical use of tasks to parallelize code that traverses a linked list. If a parallel loop were created to do so, the application would first have to ascertain the number of items in the list, so that the work associated with them could be distributed among the threads. This is not only cumbersome for the programmer, it also consumes execution time unnecessarily. Here, one of the threads will execute the single region and generate a task for the current node. It will then move on to the next node and generate a task for it. Other threads may immediately start executing tasks that are ready. If too many tasks are created, the implementation is permitted to suspend the generation of tasks so that some of them may be executed before more are created.

Since each task will perform work related to the node that was current at the time of execution, the value of the variable `currentNode` must also be saved for retrieval when the work is performed. For this reason, variables passed to a task are firstprivate by default.

```
/***********************************************************
Explicit OpenMP tasks to parallelize code traversing a
linked list in C. Though currentNode is firstprivate by
default, it is good practice to use explicit data
scoping attributes.
***********************************************************/
void processList(Node * list)
{
    #pragma omp parallel
     #pragma omp single
      {
        Node * currentNode = list;
        while (currentNode) {
         #pragma omp task firstprivate(currentNode)
          doWork(currentNode);
              currentNode = currentNode->next;
      }
     }
    }
```

Example 6.6
OpenMP task directive.

6.2.4   Synchronization

When a new value is computed for a shared variable during program execution, that value is initially stored either in cache memory or in a register. In either case, it may be accessible only to the thread that performed the operation. At some subsequent, unspecified time, the value is written back to main memory from which point on it can be read by other threads. The time when that occurs will depend partly on the way in which the program was translated by the compiler and in part on the policies of the hardware and operating system. Since the programmer needs some assurances on the availability of values so that threads may cooperate to complete their work, OpenMP provides its own set of rules on when new values of shared data must be made available to all threads and provides several constructs to enforce them.

Many OpenMP programs rely on the barriers that are performed at the end of work-sharing regions in order to ensure that new values of shared data are available to all threads. While this is often sufficient to synchronize the actions of threads, there is also a `barrier` directive that may be explicitly inserted into the code if needed.

It is then essential that all the threads in the current team execute the barrier, since otherwise the subset of threads that do so will wait for the remaining threads indefinitely, leading to deadlock. Note that all waiting tasks will be processed at a barrier, so that the program can safely assume that this work has also been completed when the threads proceed past this synchronization point in the program.

There is also a directive that can be used to enforce completion of the execution of waiting tasks. Whereas a barrier will require all threads to wait for the completion of all tasks that have been created, both implicit and explicit, the `taskwait` directive applies only to specific explicit tasks. The `taskwait` directive stipulates that the currently executing parent task wait for the completion of any child tasks before continuing. In other words, it applies only to child tasks created and not to any subsequently generated tasks.

Many computations do not require that shared data accesses occur in a specific order, but do need to be sure that the actions of threads are synchronized to prevent data corruption. For example, if two different threads must modify the value of a shared variable, the order in which they do so might be unimportant. But it is necessary to ensure that both updates do not occur simultaneously. Without some form of protection, it is conceivable that a pair of threads both access the variable being modified in quick succession, each obtaining the same "old" value. The two threads could then each compute their new value and write it back. In this last step, the value of the first update will simply be overwritten by the value produced by the slower thread. In this manner, one of the updates is simply lost. To prevent this from happening, the programmer needs a guarantee that the corresponding computations are performed by just one thread at a time. In order to ensure this mutual exclusion, the OpenMP programmer may use the `critical` directive to ensure the corresponding critical region may be executed by only one thread at a time. When a thread reaches such a region of code at runtime, it must therefore first check whether another thread is currently working on the code. If this is the case, it must wait until the thread has finished. Otherwise, it may immediately proceed. A name may be associated with a critical construct. If several critical directives appear in the code with the same name, then the mutual exclusion property applies to all of them. This means that only one thread at a time may perform the work in any critical region with a given name.

Locks offer an alternative means of specifying mutual exclusion. They are discussed briefly in Section 6.2.5. Note that for some simple cases, the `atomic` directive suffices. For example, if the operation that must be protected in this fashion is a simple addition or subtraction of a value, then prefixing this by the `atomic` directive ensures that the fetching, updating, and writing back of the modified value occurs as if it were a single indivisible operation. The value must also be flushed back to memory so that another thread will access the new value when it performs the operation.

One of the least understood features of the OpenMP API is the `flush` directive, which has the purpose of updating the values of shared variables for the thread that encounters it. In contrast to the features introduced earlier, this directive does not synchronize the actions of multiple threads. Rather, it simply ensures that the thread executing it will write any new shared values that are locally stored back to shared memory and will retrieve any new values from the shared memory for shared data that it is using. However, if two different threads encounter a flush directive at different times, then they will not share their new values as a result. The first of the threads will make its data available but will not have access to those created by the thread that has not yet reached that point in the program. As a result, it can be tricky to synchronize the actions of multiple threads via flushing, although it can offer an efficient way of doing so once the principle is understood.

Most existing shared memory systems provide support for cache coherency, which means that once a cache line is updated by a thread, any other copies of that line (including the corresponding line in main memory) are flagged as being "dirty" (Bircsak et al. 2000). This means that the remaining threads know that the data on other copies is invalid. As a result, any attempt to read data from that line will result in the updated line being shared. In other words, the thread will indeed get the new values—in contrast to the description mentioned earlier. Thus, on many of today's systems, regular flushing has little if any impact on performance. As soon as data is saved in cache, the system will ensure that values are shared. (So there is only a problem if it is still in a register.) Some performance problems result from this, as discussed in the following. Also, not all systems guarantee this and, in the future, it is likely to be a bigger problem, as it can be expensive to support cache coherency across large numbers of threads.

6.2.4.1 Performing Reductions in OpenMP

The following example illustrates two different ways in which the sum of the elements of an array may be computed by a team of threads. In the first case, each thread has a private copy of the variable `local_sum` (which has been initialized to immediately prior to the parallel region). The default loop schedule will be applied to assign iterations to threads. Each iteration will update the local sum. In order to combine these values, the global sum must be formed. Since it is necessary to ensure that the threads do not interfere with each other while adding their own local sum to the global value, the application developer must protect this update. Here, a critical region encloses the corresponding operation.

The second part of this program makes use of the OpenMP reduction facility to perform the same computation. Here, the variable that will hold the global sum is initialized, and then a parallel region is started to carry out the work. The reduction clause is used to indicate that there is a reduction operation, but also the kind and the variable that will hold the result. The rest of the work of deciding how best to perform the reduction operation is left up to the implementation, which is free to choose the most efficient strategy it can find for combining the partial results.

```
/ * * * * * * * * * * * * * * * * * * * * * * * * * * * * * * * * * * * * * * * * * * * * * * * * * * * *
These two loops achieve the same end. The first shows a
reduction operation using an OpenMP for loop and a
critical region, without which data corruption might
otherwise occur. The second uses an OpenMP for loop with
```

a reduction clause to direct the implementation to take
care of the reduction.

```
*********************************************************/
#pragma omp parallel shared(array,sum)
firstprivate(local_sum)
{
  #pragma omp for private(i,j)
    for(i = 0; i < max_i; i++){
      for(j = 0; j < max_j; ++j)
        local_sum += array[i][j];
}
#pragma omp critical
sum += local_sum;

}   /*end parallel*/
/*** Alternate version ***/
sum = 0;
#pragma omp parallel shared(array)
{
    #pragma omp for reduction(+:sum) private(i,j)
      for(i = 0; i < max_i; i++){
        for(j = 0; j < max_j; ++j){
          sum += array[i][j];
      }
    }
} /*end parallel*/
```

Example 6.7
Reductions in OpenMP.

6.2.5   OpenMP Library Routines and Environment Variables

In addition to directives, the API provides a set of library calls that may be used
to set, retrieve and get information on some important execution parameters. These
include the number of threads that will be used to execute the next parallel region,
the schedule to be applied to parallel loops that have been declared with a runtime
schedule, and the maximum permitted level of nested parallelism (Blikberg and Sørevik 2005). Further, it is possible to find out whether the number of threads that will
execute a parallel region may be dynamically adjusted, whether nested parallelism is
enabled and to modify these aspects of program behavior. Each member of the team is
numbered consecutively with a nonnegative integer, whereby the master is assigned
the value 0. Most importantly, a thread is able to retrieve its own thread ID. It can find
out whether it is currently executing a parallel region, how many parallel regions it
is contained in, the number of threads in the team executing any of these levels, the

thread ID of its ancestor in any of those levels, and the maximum number of threads available to a program.

Many of these values can be set prior to execution by assigning a value to the corresponding environment variable. Environment variables can be also used to set the stack size and a wait policy. We discuss these briefly in Section 6.3.

Another set of routines is available for the explicit use of locks to control threads' access to a region of code. In addition to providing the same kind of control as a critical construct, it is possible to test the value of a lock prior to an attempt to acquire it. If the test shows that it is set, the thread may be able to perform other work rather than simply waiting. The OpenMP library should be included in C/C++ programs as a preprocessor directive with #include <omp.h>.

6.2.5.1 SPMD Programming Style

It is even possible to rely primarily on the library routines in order to distribute the work of a program to threads. Since this involves the explicit mapping of work to the individual threads, it is often considered to be a rather low-level style of programming. A strength of OpenMP is that directives may be used where appropriate and, in the same program, this lower-level style of coding adopted where precise control of the mapping of work is necessary.

```
/ * * * * * * * * * * * * * * * * * * * * * * * * * * * * * * * * * * * * * * * * * * * * * * * * * * *
SPMD code style using threadids and using a barrier
* * * * * * * * * * * * * * * * * * * * * * * * * * * * * * * * * * * * * * * * * * * * * * * * * * * * /
#pragma omp parallel private(my_id, local_n) shared(a,p)
{
 p = omp_get_num_threads();
 my_id = omp_get_thread_num();

if (my_id == 0){
 printf("I am the master of %d thread(s).\n", p);
}else{\{}
  printf("I am worker #%d\n", my_id);
}
if (my_id == 0){
 a = 11;
local_n = 33;
}else{
 local_n = 77;
}
#pragma omp barrier
if (my_id != 0){
 local_n = a*my_id;
}
```

```
    #pragma omp barrier
    printf("Thread %d has local_n = %d\n", my_id, local_n);
    }
    printf("a = %d\n", a);
```

Example 6.8
Explicit assignment of work to threads in SPMD style.

6.3   Implementation of OpenMP

In contrast to some other high-level programming languages for parallel computing, the implementation of OpenMP is relatively straightforward. It typically involves
two components, an OpenMP-aware compiler that is able to recognize and appropri-
ately process the OpenMP constructs supplied by the application developer, and the
compiler's runtime library, which is responsible for creating and managing threads
during a program's execution (Liao et al. 2006). The routines of a runtime library are
specific to a given compiler. Although there is a typical strategy followed by most
compilers, their details differ somewhat and there are no standards with respect to
either the functionality or the name of runtime library routines.

Today, most compilers that implement one or more of the base languages are also
able to translate OpenMP. When such a compiler translates a user-supplied Fortran,
C, or C++ code, it will also encounter any OpenMP directives or user-level library
calls that have been inserted into the program. Assuming that any compiler options
required for OpenMP translation have been specified, it will translate the OpenMP
constructs by suitably adapting the code and inserting calls to its runtime library routines. Otherwise the OpenMP constructs will be ignored.

The implementation of the parallel directive is the most fundamental step in the
translation process. Typically, the code that is lexically contained within the parallel
construct is converted into a procedure, an approach that is known as outlining. References to shared variables in the region are replaced by pointers to the memory location
for the shared object and passed to the routine as an argument while private variables
are simply treated as being local to this routine. A minor adaptation of this strategy
serves to realize firstprivate variables; their initial value is passed to the procedure,
typically in the form of an additional parameter. Then a runtime library routine is
invoked to fork the required number of threads and to pass the outlined procedure to
them for execution. In practice, implementations avoid the repetitive creation and termination of threads at the start and end of parallel regions by allowing threads other
than the master to sleep between different parallel regions. The latest standard permits the user to help determine whether threads will busy-wait or sleep at other places
where they are idle, such as when they are waiting to enter a critical region. If wait
times are likely to be short, busy-waiting is often preferred since they will quickly
respond to a changed situation. However, this can consume resources that may be

needed by active threads. So when the wait can be longer, it might be preferable to put them to sleep.

A parallel loop will typically be replaced by, first, a routine that each thread invokes to determine the set of loop iterations assigned to it. Then the iterations are performed and a barrier routine invoked. For static schedule kinds, each thread may independently determine its share of the work. If a dynamic or guided schedule has been specified, then a thread may need to carry out multiple sets of iterations. In such cases, once it has completed one set, it will again invoke a routine to get another portion of the work. In this case, more coordination is required behind the scenes and the corresponding routine is involved several times. As a result, overheads are slightly higher. However, if the loop iterations contain varying amounts of work, it is often faster than a static schedule. Parallel sections are often converted to a parallel loop that has as many iterations as there are sections. Branches to the individual sections ensure that the ith iteration will perform exactly the work of the ith section.

A `single` directive is typically executed by the first thread that reaches it. To accomplish this, each thread will test the value of a shared variable that indicates whether or not the code has been entered. The first thread to reach it will set the variable and thus ensure that other threads do not attempt to perform the computations. As with any access to a shared variable, this test and set should be done atomically by the implementation.

There are several ways in which OpenMP tasks may be implemented (Addison et al. 2009). A simple strategy is to create a queue of tasks, which holds all tasks that have been generated but whose execution has not been completed. More sophisticated strategies may implement one or more queues of tasks for each thread and enable threads to steal tasks from other threads' queues if they have no more tasks in their own queue (Frigo et al. 1998, Duran et al. 2008). This might, for instance, enable some amount of data locality. Since tied tasks may be suspended, and thus put back onto a queue, there might be an additional queue per thread to hold these. Such tasks cannot be stolen by other threads so this will separate them from those that could be executed by any thread. Implementations must also consider whether to prefer to continue to generate tasks until a synchronization point is encountered or some threshold has been reached (which may depend on the overall number of tasks generated, the number already in a queue, or some additional criteria), or whether they should use some other strategy to decide when to start executing tasks that are ready.

Critical regions are usually implemented with low-level locks managed by the runtime. Threads that are waiting to enter a critical region may do so actively, i.e., by frequently checking to find out if the region is still being worked on. They can also do so passively, in which case the thread simply sleeps and is woken up when the computation is ready for it to proceed. Both modes can typically be supported by an implementation. Whereas the former case usually allows the thread to begin its work faster, it also causes the thread to use resources that may be needed by other threads, thereby interfering with their progress. Nevertheless, if the wait time is expected to be very short, it may be a good choice. If the waiting time is long, the latter may be better. Since the implementation may not be able to determine which policy is preferable, the OpenMP API allows the application developer to influence its strategy.

The OpenMP (user-level) library routines are often simply replaced by a corresponding runtime procedure. Since some of runtime routines are frequently invoked—this applies particularly to the barrier implementation and to the function that retrieves a thread's identifier—they are typically very carefully crafted. There are, for instance, a variety of algorithms for performing barriers and efficient versions can significantly outperform naïve barrier implementations (Nanjegowda et al. 2009). Locks and atomic updates might also have straightforward implementations. The underlying thread package chosen for thread management operations will depend on the options available. Where portability is required, Pthreads is often selected. But most systems provide more efficient alternatives. Note that OpenMP is beginning to be implemented on some kinds of systems that do not provide a fully fledged operating system and thus sometimes have minimal support for thread operations. For these systems too, the implementation will choose the best primitives available (Chapman et al. 2009).

Although OpenMP has most often been implemented on cache-coherent shared memory parallel computers, including multicore platforms, it has also been implemented on non-cache-coherent platforms, large distributed shared memory machines, and there are a few implementations for distributed memory machines also (Huang et al. 2003, Marowka et al. 2004, Hoeflinger 2006). OpenMP has begun to be used as a programming model for heterogeneous multicore architectures (Liu and Chaudhary 2003, Chapman et al. 2009), where different kinds of cores are tightly coupled on a chip or board. An early implementation of this kind targeted the Cell, which combines a general purpose core (PPE) with multiple special purpose cores (SPEs); the SPEs do not share memory with the PPE, which makes the translation considerably tougher (O'Brien et al. 2008). A similar approach was taken to enable the convenient use of ClearSpeed's accelerators in conjunction with general purpose multicore hardware (Gaster and Bradley 2007). Given the growth in platforms that provide some kind of accelerator and need to facilitate programming across such systems, we expect to see more attempts to provide such implementations in the future (Ayguadé et al. 2009a).

Our example gives a flavor of the code that is generated to execute the loop in Example 6.7. It begins with the outlined procedure that encapsulates the work of the parallel region. The compiler has generated a name for the procedure that indicates that it is the second parallel region of the main procedure. Here, too, there are no standards: in some tools in the programming environment, these names may be made visible to the user. Next the compiler has generated a private variable in which each thread will store its local portion of the reduction operation. It then saves the original bounds of the loop that will be distributed among threads before passing these as arguments to the procedure that will determine static schedule. Note that each thread independently invokes this routine, and that it will use the bounds as well as the thread number to determine its own set of iterations. A call to the barrier routine has been inserted to ensure that the threads wait until all have completed their share of the parallel loop. Next, a critical region is used to combine the local sums, in order to complete the reduction. Note that reduction operations are another feature that permits a variety of implementations, some of which are much more efficient than others.

```
/************************************************************
Possible implementation of the code in the 2nd parallel
region in Example 6.7
************************************************************/
static void __ompregion_main2(thrdid)
 /* var declarations omitted */
 local_sum = 0;
 limit = max_i + -1;
 do_upper = limit;
 do_lower = 0;
 last_iter = 0;
__ompc_static_init(thrdid, 2, &do_lower,
           &do_upper, &do_stride, 1, 1);
 if(do_upper > limit)
  {
   do_upper = limit;
  }
 for(local_i = do_lower; local_i <= do_upper;
  local_i = local_i + 1)
  {
   local_j = 0;
   while(local_j < max_j)
  {
   local_sum = array[local_i][local_j] + local_sum;
   local_j = local_j + 1;
  }
 }
__ompc_reduction(thrdid, &lock);
sum = local_sum + sum;
__ompc_end_reduction(thrdid, &lock);
__ompc_barrier();
return;
} /* __ompregion_main2 */
```

Example 6.9
OpenMP implementation strategy.

6.4 Programming for Performance

It is hard to overemphasize the importance of carrying out some amount of performance tuning once an initial, correct parallel program has been created. Many application developers are not aware of the potential for enhancing the overall runtime behavior if certain aspects of the parallel code are overlooked. The performance of parallel programs relies on adequately addressing that matter of ensuring efficiency in how the program will use its available resources, primarily the memory hierarchy and processor threads (Jin et al. 2007). A naïve approach to the use of shared variables often leads to programs with unexpected behavior; unfortunately it may also lead to code containing obscure bugs. However, an informed approach will prevent the introduction of errors and lead to better performance. Once an optimal sequential version of the code is prepared it can be incrementally parallelized and tuned for better performance.

Ideally, all of the available threads should continuously perform useful work throughout the execution of a code. In the OpenMP context, the programmer should attempt to ensure that all major computations are parallelized. However, the extent to which the code is parallelized may depend on the overall performance needed, and it is quite possible that there may be remaining sequential parts, e.g., to initialize the code. Now that we have discussed the basic strategy adopted to implement OpenMP, we can use this insight to now look into how unnecessary overheads can be avoided.

Use of the `parallel` directive deserves careful consideration; if improperly used it may actually cause performance degradation. This construct entails the overhead necessary for the creation (or waking up) of thread teams, so it should be used infrequently and enclose as much code as possible. For instance, successive `parallel for` directives should be written as one `parallel` region enclosing successive `for` directives. An important potential benefit of this is that cache utilization may be optimized over a larger region of code: if the different loops access the same data, the performance advantage of merging these parallel regions may be significant. The overall goal should be to maximize the size and minimize the frequency of `parallel` directives. It is also prudent to use the `if` clause for parallel and task regions where the amount of parallelism may be not be known at compile time. By providing a condition to test for sufficient concurrency at runtime, unnecessary overheads can be avoided.

Any synchronization implies that threads may have to wait for another thread to complete an operation (Weng and Chapman 2003). The longer the wait, the more likely it is to degrade overall program performance. The OpenMP programmer should attempt to eliminate unnecessary synchronization and avoid placing unnecessary computations into critical regions or code portions that are controlled by locks. Avoiding unnecessary barriers requires knowing which constructs entail implicit barriers and which do not. Successive implicit barriers may not be obvious but can be eliminated with strategic use of a `nowait` clause. Limiting the use of `taskwait` should also be considered. Task parallelism often involves great numbers of tasks which should not be kept waiting unnecessarily.

The overheads of synchronization are so important that we have discussed them separately. However, there are many other overheads that arise when OpenMP programs are executed. Any time each thread performs operations that were not part of the original program, overheads are incurred. This is also the case if some work is replicated, since there is no immediate performance benefit obtained by it.

The manner in which memory is used by a program has a great impact on performance, namely in its use of the cache (Jin et al. 1999). As mentioned previously, cache lines are flagged in their entirety when "dirty." It is not just the needed variable that is updated but the entire cache line that contains it! This is expensive! Therefore it is prudent to minimize cache misses whenever possible. One way to avoid this is to make sure loops are accessing array elements in the proper order, either by row or column, for the language being used. Fortran arrays are stored in column-major order and C/C++ are row-major, so loops should be modified if needed to access array elements accordingly.

One side effect of cache coherent systems is false sharing. This is the interference among threads writing to different areas of the same cache line. The write from one thread causes the system to notify other caches that this line has been modified, and even though another thread may be using different data it will be delayed while the caches are updated. False sharing often prevents programs from scaling to a high number of threads. It may require careful scrutiny of the program's access patterns if it is to be avoided.

It is also important to achieve a balanced workload. The time required for a thread to carry out its portion of the computation should be, as far as possible, the same as the time required for each of the other threads between any pair of synchronization points. This is generally accomplished with a suitable loop schedule for the given algorithm. Some experimentation may be necessary to determine the schedule that provides the best performance for any given use of a loop directive.

OpenMP provides an easy model for incrementally writing parallel code, and it is particularly easy to obtain an initial parallel program, but special considerations are needed to ensure that a program performs well. With due diligence, an OpenMP programmer can address these common performance obstacles and obtain a parallel application that is both efficient and scalable.

6.5 Summary

With the rapid proliferation of multicore platforms and the growth in the number of threads that they may support, there is an urgent need for a convenient means to express the parallelism in a broad variety of applications. OpenMP began as a vehicle for parallelizing (primarily) technical computations to run on small shared memory platforms. Since that time, it has evolved in order to provide a means to express a number of parallel programming patterns that can be found in modern computations and to support the parallelization of applications written in Fortran, C, and C++. The latest version provides support for the expression of multilevel parallelism, for loop and task-based parallelism, for dynamic adjustments of the manner of the program's execution (including the ability to modify the number of threads that will be used to execute a parallel region), includes features for fine-grained load balancing and offers ability to write high level, directive-based code as well as low level code that specifies the instructions that are to be executed by the different threads explicitly. The growth in terms of numbers of features has been relatively modest.

The challenges for OpenMP are therefore to support the programming of systems with large numbers of threads and to ensure that it is able to express the diverse patterns of parallelism that occur in modern technical and nontechnical application codes alike. Current work is exploring means to provide for a coordinated mapping of work and data to threads, to enable error handling, and to enhance the task interface. With the introduction of systems based on heterogeneous cores, the complexity of the application development process has moreover once more increased significantly. It will be interesting to see how well OpenMP may target such systems also. Early work has already begun to address this topic (Ayguadé et al. 2009a, Huang and Chapman 2009). The OpenMP ARB is actively considering a variety of strategies, and features, for addressing these challenges, as well as providing additional help to deal with errors and to enhance several of its existing features.

References

Addison, C., J. LaGrone, L. Huang, and B. Chapman. 2009. OpenMP 3.0 tasking implementation in OpenUH. In *Open64 Workshop in Conjunction with the International Symposium on Code Generation and Optimization*. http://www.capsl.udel.edu/conferences/open64/2009/ (accessed October 5, 2009).

Adve, S. V. and K. Gharachorloo. 1996. Shared memory consistency models: A tutorial. *Computer*, 29(12), 66–76.

Ayguadé, E., R. M. Badia, and D. Cabreral. 2009a. A proposal to extend the OpenMP tasking model for heterogeneous architectures. In *International Workshop on OpenMP*, Dresden, Germany, pp. 154–167.

Ayguadé, E., B. Blainey, A. Duran et al. 2003. Is the schedule clause really necessary in OpenMP? In *Workshop on OpenMP Applications and Tools*, Toronto, Ontario, Canada, pp. 147–159.

Ayguadé, E., N. Copty, A. Duran et al. 2009b. The design of OpenMP tasks. *IEEE Transactions on Parallel and Distributed Systems*, 20(3), 404–418.

Bircsak, J., P. Craig, R. Crowell et al. 2000. Extending OpenMP for NUMA machines. *Scientific Programming*, 8(3), 163–181.

Blikberg, R. and T. Sørevik. 2005. Load balancing and OpenMP implementation of nested parallelism. *Parallel Computing*, 31(10–12), 984–998.

Bronevetsky, G. and B. R. de Supinski. 2007. Complete formal specification of the OpenMP memory model. *International Journal of Parallel Programming*, 35(4), 335–392.

Chandra, R., L. Dagum, D. Kohr, D. Maydan, J. McDonald, and R. Menon. 2000. *Parallel Programming in OpenMP*. San Diego, CA: Morgan Kaufmann Publishers, Inc.

Chapman, B. M., L. Huang, H. Jin, G. Jost, and B. R. de Supinski. 2006. Toward enhancing OpenMP's worksharing directives. In *Euro-Par* 2006, Dresden, Germany, pp. 645–654.

Chapman, B., L. Huang, E. Stotzer et al. 2009. Implementing OpenMP on a high performance embedded multicore MPSoC. In *Proceedings of Workshop on Multithreaded Architectures and Applications (MTAAP'09)* in conjunction with *International Parallel and Distributed Processing*, Rome, Italy, pp. 1–8.

Chapman, B., G. Jost, and R. van der Pas. 2008. *Using OpenMP: Portable Shared Memory Parallel Programming*. Cambridge, MA: The MIT Press.

Compunity. 2009. cOMPunity—The community of OpenMP users. http://www.compunity.org/ (accessed October 5, 2009).

Curtis-Maury, M., X. Ding, C. D. Antonopoulos, and D. S. Nikolopoulos. 2005. An evaluation of OpenMP on current and emerging multithreaded/multicore processors. In *First International Workshop on OpenMP*, Eugene, OR.

Duran, A., J. Corbalán, and E. Ayguadé. 2008. Evaluation of OpenMP task scheduling strategies. In *Proceedings of the 4th International Workshop on OpenMP (IWOMP'08)*, Lafayette, IN, pp. 101–110.

Frigo, M., C. Leiserson, and K. H. Randall. 1998. The implementation of the Cilk-5 multithreaded language. In *SIGPLAN Conference on Programming Language Design and Implementation*, Montreal, Quebec, Canada, pp. 212–233.

Gan, G. 2009. Tile reduction: The first step towards tile aware parallelization in OpenMP. In *International Workshop on OpenMP*, Dresden, Germany, pp. 140–153.

Gaster, B. and C. Bradley. 2007. Exploiting loop-level parallelism for SIMD arrays using OpenMP. In *International Workshop on OpenMP*, Dresden, Germany, pp. 89–100.

Hoeflinger, J. P. 2006. Extending OpenMP to clusters. Technical report, Intel Inc.

Huang, L. and B. Chapman. 2009. Programming heterogeneous systems based on OpenMP. Accepted for publication in *Mini-Symposium on Programming Heterogeneous Architectures, International Conference on Parallel Computing (ParCo'09)*.

Huang, L., B. Chapman, and R. Kendall. 2003. OpenMP on distributed memory via global arrays. In *Parallel Computing*. Dresden, Germany.

Intel Corporation. 2009. Intel and Intel compiler. http://software.intel.com/en-us/intel-compilers/ (accessed October 5, 2009).

Jin, H., B. Chapman, and L. Huang. 2007. Performance evaluation of a multi-zone application in different OpenMP approaches. In *International Workshop on OpenMP*, Beijing, China.

Jin, H., M. Frumkin, and J. Yan. 1999. The OpenMP implementation of NAS parallel benchmarks and its performance. Technical Report NAS-99-011, NASA Ames Research Center.

Jost, G., J. Labarta, and J. Gimenez. 2004. What multilevel parallel programs do when you are not watching: A performance analysis case study comparing MPI/OpenMP, MLP, and nested OpenMP. In *Workshop on OpenMP Applications and Tools*, Houston, TX.

Larsen, P., S. Karlsson, and J. Madsen. 2009. Identifying inter-task communication in shared memory programming models. *International Workshop on OpenMP*, Dresden, Germany, pp. 168–182.

Liao, C., O. Hernandez, B. Chapman, W. Chen, and W. Zheng. 2006. OpenUH: An optimizing, portable OpenMP compiler. In *12th Workshop on Compilers for Parallel Computers*, La Coruna, Spain.

Liu, F. and V. Chaudhary. 2003. Extending OpenMP for heterogeneous chip multiprocessors parallel processing. In *Proceedings of International Conference on Parallel Processing*, Kaohsiung, Taiwan, pp. 161–168.

Marowka, A., Z. Liu, and B. Chapman. 2004. OpenMP-Oriented applications for distributed shared memory architectures. *Concurrency and Computation: Practice and Experience*, 16(4), 371–384.

Mattson, T. G. 2003. How good is OpenMP? *Scientific Programming*, 11(2), 81–93.

Nanjegowda, R., O. Hernandez, B. Chapman, and H. Jin. 2009. Scalability evaluation of barrier algorithms for OpenMP. *International Workshop on OpenMP*, Dresden, Germany, pp. 42–52.

O'Brien, K., K. O'Brien, Z. Sura, T. Chen, and T. Zhang. 2008. Supporting OpenMP on cell. *International Journal of Parallel Programming*, 36(3), 289–311.

Open64. 2005. The Open64 compiler. http://open64.sourceforge.net (accessed October 5, 2009).

OpenMP. 2009. OpenMP: Simple, portable, scalable SMP programming. http://www.openmp.org/ (accessed October 5, 2009).

OpenMP. ARB. 2008. OpenMP application programming interface, version 3.0. http://www.openmp.org/mp-documents/spec30.pdf (accessed October 5, 2009).

The Portland Group. 2009. OpenMP: Shared-memory parallelism for Fortran, C and C++. http://www.pgroup.com/resources/openmp.htm (accessed October 5, 2009).

Sun Microsystems, Inc. 2005. OpenMP support in Sun Studio compilers and tools. http://developers.sun.com/solaris/articles/studio_openmp.html (accessed October 5, 2009).

Weng, T.-H. and B. Chapman. 2003. Toward optimization of OpenMP codes for synchronization and data reuse. In *The 2nd Workshop on Hardware/Software Support for High Performance Scientific and Engineering Computing (SHPSEC-03)*, in conjunction with the *12th International Conference on Parallel Architectures and Compilation Techniques (PACT-03)*, New Orleans, LA.

Part III

Programming Heterogeneous Processors

Chapter 7

Scalable Manycore Computing with CUDA

Michael Garland, Vinod Grover, and Kevin Skadron

Contents

7.1   Introduction

The applications that seem most likely to benefit from major advances in computational power and drive future processor development appear increasingly throughput oriented, with products optimized more for data or task parallelism depending on their market focus (e.g., HPC vs. transactional vs. multimedia). Examples include the simulation of large physical systems, data mining, and ray tracing. *Throughput-oriented* workload design emphasizes many small cores because they eliminate most of the hardware needed to speed up the performance of an individual thread. These simple cores are then multithreaded, so that when any one thread stalls, other threads can run and every core can continue to be used to maximize the application's overall throughput. Multithreading in turn relaxes requirements for high performance on any individual thread. Small, simple cores therefore provide greater throughput per unit of chip area and greater throughput within a given power or cooling constraint. The high throughput provided by "manycore" organizations has been recognized by most major processor vendors.

To understand the implications of rapidly increasing parallelism on both hardware and software design, we believe it is most productive to look at the design of modern GPUs (Graphics Processing Units). A decade ago, GPUs were fixed-function hardware devices designed specifically to accelerate graphics APIs such as OpenGL and Direct3D. In contrast to the fixed-function devices of the past, today's GPUs are fully programmable microprocessors with general-purpose architectures. Having evolved in response to the needs of computer graphics—an application domain with tremendous inherent parallelism but increasing need for general-purpose programmability— the GPU is already a general-purpose manycore processor with greater peak performance than any other commodity processor. GPUs simply include some additional hardware that typical, general-purpose CPUs do not, mainly units such as rasterizers that accelerate the rendering of 3D polygons and texture units that accelerate filtering and blending of images. Most of these units are not needed when using the GPU as a general-purpose manycore processor, although some can be useful, such as texture caches and GPU instruction-set support for some transcendental functions. Because GPUs are general-purpose manycore processors, they are typically programmed in a fashion similar to traditional parallel programming models, with a single-program, multiple data (SPMD) model for launching a large number of concurrent threads, a unified memory, and standard synchronization mechanisms.

High-end GPUs cost just hundreds of dollars and provide teraflop performance while creating, executing, and retiring literally *billions* of parallel threads per second, exhibiting a scale of parallelism that is orders of magnitude higher than other platforms and truly embodies the manycore paradigm. GPUs are now used in a wide range of computational science and engineering applications, and are supported by several major libraries and commercial software products.

7.2   Manycore GPU Machine Model

Before programming a GPU, it is helpful to have a simple mental model of the GPU architecture. GPUs assume the presence of plentiful parallelism and hence the hardware is optimized for total throughput rather than single-threaded latency. GPUs therefore consist of many simple, highly multithreaded processing elements per chip and high memory bandwidth. Keeping the cores simple increases the total number of cores that can be integrated, and multithreading allows the GPU to efficiently consume the available memory bandwidth. Today's GPUs can execute *tens of thousands* of threads at the same time, and their programming models are designed to scale orders of magnitude beyond that.

The overall machine that we target is a *heterogeneous* platform consisting of both a host CPU and associated GPU. The CPU is a processor designed to optimize the performance of sequential codes, whereas the GPU is a processor designed to optimize

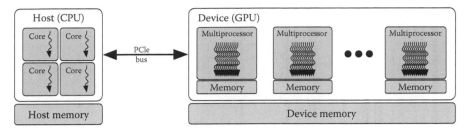

FIGURE 7.1: The heterogeneous hardware environment typical of GPU computing.

the performance of large parallel workloads. Since any particular program will consist of both latency-sensitive sequential sections and throughput-sensitive parallel sections, it is advantageous to have processors optimized for both types of workload (Figure 7.1).

From a programmer's perspective, the key ingredients of the GPU architecture can be broken into three categories. The chip itself consists of a collection of multithreaded multiprocessors called *SMs*, each of which can run a large collection of threads. The hardware provides a unified memory model with fast local on-chip memories and global off-chip memory visible to all threads. These memory spaces provide a relaxed memory-consistency model [1], meaning that memory operations from one thread might appear to other threads out of order. Thread barrier, memory fence, and atomic read-modify-write primitives are provided to provide synchronization and ordering guarantees where those are required.

Each of the multiprocessors in the GPU is designed to manage and execute a large population of threads—up to 1536 in current generation hardware. Every thread represents an independent execution trace in that it possesses its own program counter, stack, and scalar register set. Threads communicate through shared memory spaces and synchronize at barriers. In short, they are fundamentally similar to user-level CPU threads.

In reality, a multiprocessor will contain far fewer physical processing elements than the total number of threads it can support. Therefore, the virtual processors represented by the threads are time multiplexed in a fine-grained fashion onto the physical processing elements, with a new thread running every cycle. Each scalar processor hosts dozens of threads, providing high latency tolerance (e.g., for accesses to off-chip memory). Correct thread execution does not depend on which processing element hosts the thread's virtual processor.

The processing cores themselves support a fully general-purpose, scalar instruction set. They provide full support for both integer and IEEE floating point—at both single and double precision—arithmetic. They provide a standard load/store model of memory, which is organized as a linear sequence of bytes. And they support the various other features typical of modern processors, including normal branching semantics, virtual memory and pointers, function calls, etc. The GPU also offers some instruction set support for important transcendental functions such as trigonometric functions and reciprocal square root that other commodity architectures do not. These can dramatically improve performance over conventional ISAs and are important across a wide range of applications.

7.3   Structure of CUDA Programs

CUDA provides an environment for writing parallel programs that run on heterogeneous machines. These heterogeneous machines are assumed to consist of one or more host processors and one or more compute devices. A *host* processor is typically a traditional CPU that is optimized for sequential execution of a single thread. A *device* processor compute device is expected to be a processor such as a GPU that is optimized for parallel execution of many thousands of threads.

A CUDA programmer writes a single program in a standard sequential language, augmented with minimal extensions to describe parallelism. Throughout this chapter we will describe the CUDA programming model in terms of the CUDA C/C++ language, but it is important to emphasize that the CUDA architecture can support essentially any language choice. Both CUDA C/C++ and CUDA Fortran compilers are available today, and for all practical purposes, any traditional language used for writing sequential programs can be compiled for the CUDA platform.

7.3.1   Program Placement

The host and device processors will in general be separate physical chips with potentially separate memory spaces. Therefore, portions of a CUDA program will consist of data that resides on the device and functions that run on the device. We refer to these as as *device data* and the *device program*, respectively. Similarly, we refer to the data that lives in host memory and the part of the program that runs on the host as the *host data* and *host program*, respectively. A CUDA program is the combination of its constituent host and device programs. While CUDA provides support for multithreaded host programs working with multiple separate devices, we will focus on the simpler case of a single-threaded sequential host program working with a single parallel device.

The programmer declares the placement of variables, either on the host or device, using declaration modifiers as in this example:

```
__device__  int x;  // x resides in device memory
__host__    int y;  // y resides in host memory
```

This program fragment specifies that integer variables x and y are to be allocated in the GPU and CPU memory, respectively.

Functions are also annotated to indicate where they may execute.

```
__device__ int device_function(int *p) { ... }

__host__ int host_function(int *p) { ... }

__host__ __device__ int function(int *p) { ... }
```

A device function may only be called within the device program and will execute on the device. A host function may only be called within the host program and will execute on the host. Functions marked with both specifiers may be called in either the host or device program and will execute on the processor where they are called. Functions without any placement annotation are assumed to be host functions.

7.3.2 Parallel Kernels

A CUDA program begins its execution as a standard sequential program on the host. The host program can initiate computations on the device by invoking a *kernel*: a parallel computation that executes a single program across many parallel threads. Kernel functions, which may be used as the program entry point for a kernel, are indicated with a special __global__ declaration modifier. They are launched with a function call-like syntax as in this example code fragment:

```
// Kernel entry point.  Every kernel thread
// will execute the contents of this procedure.
__global__ void kernel(int *p, int n)
{
    ...
}


__host__ void launch(int *p, int n)
{
    ...
    kernel<<<nblocks, nthreads>>>(p, n);
    ...
}
```

Kernel functions may only be called from host functions. They may in turn call only device functions. Kernels are not permitted to return values and therefore have a void return type.

The programmer organizes a kernel into a grid of *thread blocks* each of which consists of a fixed number of parallel threads. This organization of threads is specified when launching the kernel, as seen in the preceding code fragment. The launch syntax f<<<P,B>>>() will launch P blocks of B threads each, for a total of PB parallel threads across the entire kernel.

Every thread block is given a unique integral coordinate that is accessible within the device program via the special variable blockIdx. This has components blockIdx.x and blockIdx.y, since the programmer may choose to launch a kernel as either a 1D or 2D grid of blocks. The dimensions of the grid are provided by the variables gridDim.x and gridDim.y, respectively. The threads of a given block are also given unique coordinates, whose components are accessible via threadIdx.x, threadIdx.y, and threadIdx.z. The dimensions of the block are available in the variables blockDim.x, blockDim.y, and blockDim.z, respectively. The runtime values of the threadIdx and blockIdx components are the consecutive integers ranging from zero up to one less than corresponding dimension. All of these special variables are set by the runtime environment and cannot be changed by the program.

The threads of a kernel start executing at the same entry point, namely, the kernel function. However, they do not subsequently need to follow the same code sequence. In particular, most threads will likely make different decisions or access different memory locations based on their unique thread/block coordinates. This style of execution is often referred to as "single program, multiple data" or SPMD. By default, when a host program launches multiple kernels in sequence, there will be an implicit barrier between these kernels. In other words, no thread of the second kernel may be launched until all threads of the first have completed. CUDA provides additional APIs for launching independent kernels whose threads may be potentially overlapped, both with other kernels and memory transfers.

All threads of a kernel have their own local variables, which are typically stored in GPU registers. These are private and are not accessible to any other thread. Threads also have direct access to any data placed in device memory. This data is common to all threads. When accessing device memory, the programmer must either use the unique thread/block coordinates to guarantee that all threads are accessing separate data, or use appropriate synchronization mechanisms to avoid race conditions between threads.

7.3.3   Communicating within Blocks

Thread blocks provide a granularity at which threads can efficiently cooperate among themselves. All threads of a kernel have access to data in device memory and all threads effectively synchronize at the end of a kernel. These are relatively expensive operations. In contrast, threads within a single block are able to communicate very efficiently.

Each block has an associated shared memory space that is private to the block but visible to all of its threads. Accesses to this memory space are also very fast since, unlike device memory that resides in external DRAM, this memory space resides on-chip. Variables within a device function are declared to live in this space with a declaration modifier:

```
__shared__ int z;  // z resides in on-chip shared memory
```

Each thread block will have its own private copy of z, which will be allocated in the on-chip shared memory when the thread block is launched.

Shared memory is much faster than global memory access—by roughly two orders of magnitude on current hardware. Consequently, device programs often copy data into shared memory, process it there, and then write results back to global memory. In this case, shared memory functions like a local scratchpad memory, and making good use of it can lead to large performance gains in practice.

Threads within a thread block run in parallel and can share data via shared memory. They may also synchronize using an explicit barrier primitive, which is exposed in CUDA C as a __syncthreads() function call. This brings all threads to a common execution point and also ensures that all outstanding memory operations have completed. The barrier itself is implemented directly in hardware and is extremely efficient. On current hardware it compiles to a single instruction; thus, there is essentially no overhead and the cost of the barrier is simply the time required for all threads to physically reach the barrier.

It is the programmer's responsibility to ensure that all threads will eventually execute the same barrier. Barriers at textually distinct positions within the program are considered different barriers. The program behavior is undefined if some threads execute a barrier and others either skip that barrier or execute a textually distinct barrier. A barrier within conditional code, such as:

```
if( P ) {  ....; __syncthreads(); ....; }
else     {  ....; __syncthreads(); ....; }
```

is only well-defined if every thread of the block takes the same branch of the conditional. Similarly, a loop containing a barrier:

```
while( P ) {  ....; __syncthreads(); ....; }
```

is only safe if every thread evaluates the predicate P in the same way.

CUDA's block-oriented programming model allows a large variety of applications to be written by allowing threads to cooperate closely within a thread block, while also allowing a set of thread blocks to work independently and coordinate across kernel launches. This basic model is designed to allow very efficient low-level programs to be written for the GPU while allowing higher level application frameworks to be built on top using the abstraction features of C and C++ language, e.g., functions and templates.

7.3.4 Device Memory Management

In addition to launching parallel kernels, the host program is also responsible for managing device memory. It can initiate bulk data transfers from the host system's main memory to the GPU's on-board memory, launch a kernel, and copy results back. Computation can be overlapped with transfer of data that is not currently needed. While explicit data transfer is not strictly necessary, because portions of host memory can be directly mapped into the GPU's address space, it can be more efficient, especially for large blocks of data.

CUDA provides a number of interfaces for managing device in different ways and for different use cases. Among these, the most basic interface is provided by the following routines:

- cudaMalloc: Allocates device memory.

```
float *dA;
cudaMalloc((void**) &dA, N * sizeof(float));
```

In this example, the host variable dA will hold the address of the allocated memory in the device address space. It may be passed as an argument to host and kernel functions like any other parameter, but may only be dereferenced on the device.

- cudaMemcpy: Performs a data transfer between the host and the device.

```
float *hA = ...,   *dA = ...;
cudaMemcpy(dA, hA, N*sizeof(float),
            cudaMemcpyHostToDevice) );
```

This example copies N floating point values starting at address hA in the host memory to the address dA in device memory.

- cudaFree: Deallocates device memory allocated with cudaMalloc.

```
cudaFree(dA);
```

7.3.5   Complete CUDA Example

Figure 7.2 shows a simple CUDA C program. It performs the parallel vector addition $C = A + B$ where A, B, and C are all 1-dimensional vectors of n floating point numbers. This program fragment assumes that the initial arrays are already defined in host memory (hA, hB, hC) and that their length (n) is known. It proceeds to allocate device memory (dA, dB, dC) of the appropriate size, and then copies data from host to device memories. Finally, it launches the kernel using a grid of $\lceil n/256 \rceil$ thread blocks. The result of the computation is left in device memory (dC) for the use of subsequent kernels. To inspect the results on the host, the program uses cudaMemcpy to transfer the data back to the host memory space. The memory allocated with cudaMalloc can also be explicitly deallocated with cudaFree when those arrays are no longer needed and the space must be reclaimed before the end of the program.

7.4   Execution of Kernels on the GPU

A running CUDA program consists of one or more host threads that can launch parallel device kernels. These host threads are executed on the CPU(s) present in the machine and may be managed with OpenMP, MPI, or another multithreading layer of the programmer's choosing.

When a host thread launches a kernel kernelA<<<P,B>>>, as illustrated in Figure 7.3, the launch request is received by the CUDA runtime layer. It is responsible for delivering this request to the hardware. Once the launch request is received by the runtime, the launching host thread is allowed to continue executing. In other words, all kernels launched by a given host thread will run asynchronously with that host thread. To synchronize with the kernels it has launched, a host thread calls one of a number of synchronizing operations, the simplest of which is cudaThreadSynchronize(). This acts as a barrier on all outstanding device activities; all kernels called prior to

```
// Compute C = A + B for n-vectors A, B, and C.
// Each thread performs one pair-wise addition
__global__ void vecAdd(int n, float* A, float* B, float* C)
{
    int i = threadIdx.x + blockDim.x * blockIdx.x;

    if( i<n )  C[i] = A[i] + B[i];
}

int main()
{
    // The n-vectors A,B are in host (CPU) memory
    float *hA = ...,   *hB = ...,   *hC = ...;
    int n = ...;

    // Allocate device (GPU) memory
    int nbytes = n * sizeof(float);
    float *dA, *dB, *dC;
    cudaMalloc((void**) &dA, nbytes);
    cudaMalloc((void**) &dB, nbytes);
    cudaMalloc((void**) &dC, nbytes);

    // Copy host memory to device
    cudaMemcpy(dA, hA, nbytes, cudaMemcpyHostToDevice);
    cudaMemcpy(dB, hB, nbytes, cudaMemcpyHostToDevice);

    // Execute the kernel on ⌈n/256⌉ blocks of 256 threads each
    int blocks = (n + 255) / 256;
    vecAdd<<<blocks, 256>>>(n, dA, dB, dC);

    // Copy results back to the host
    cudaMemcpy(hC, dC, nbytes, cudaMemcpyDeviceToHost);

    return 0;
}
```

FIGURE 7.2: A simple CUDA program that adds two n-vectors.

the barrier must complete before the host thread is allowed to proceed. Performing any memory transfer, such as by calling `cudaMemcpy()`, implicitly introduces a barrier that waits for all previous kernels to complete.

7.4.1 Kernel Scheduling

On NVIDIA GPUs, the GPU hardware is responsible for scheduling kernel initiation, thread block initiation, thread execution, and completion. The CUDA runtime layer delivers kernels to the hardware in the order they are launched. Once a kernel

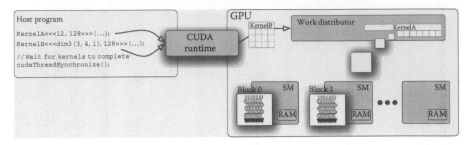

FIGURE 7.3:   As a host program launches kernels, these are delivered to the hardware which schedules blocks onto the SMs.

is scheduled for execution, a description of its launch configuration is passed to a work distributor unit. The information needed to launch a kernel includes the address of the first instruction of the kernel program, the grid dimension P, the thread block dimension B, and the resource requirements needed to run the kernel.

The work distributor will launch thread blocks from the grid onto multiprocessors as long as there are resources available to run them. These resources include thread execution contexts for each thread of the block, registers for each thread, and any space in the on-chip shared memory requested by the program. If insufficient resources are available—all multiprocessors may have already reached their limit for resident threads, for instance—the work distributor will hold further thread blocks until resources become available.

CUDA programs cannot rely on the thread blocks of a grid being launched in any particular order. Furthermore, they must assume that thread blocks, once launched, are run to completion without preemption. Hence, a kernel's thread blocks should be *independent*: the kernel should run correctly for any possible interleaving of thread block executions, sequentially or in parallel. At one extreme, this means that a kernel should execute correctly if all of its thread blocks are simultaneously running in parallel, as well as at the other extreme where blocks run one at a time in any order.

This independence restriction allows CUDA to virtualize the number of physical multiprocessors available on the target device. At the program level, we can think of a single thread block as a virtualized multiprocessor, much as we can think of a single host thread as a virtualized sequential core. A thread block encompasses some number of parallel threads, an associated memory private to its constituent threads, and their associated state. We would like a single CUDA program to scale smoothly across a wide range of physical processors; this is particularly important since the number of multiprocessors in currently deployed CUDA-capable GPUs varies by an order of magnitude. One option for virtualizing physical multiprocessors would be to time-multiplex thread blocks among them, much as threads are scheduled on multicore CPUs. However, this would in general require context swapping the state of thread blocks to and from external memory. It is common for a high-end GPU to have hundreds of thread blocks, each with a hundred or more threads, running at any one time. Frequently transferring the corresponding processor state to and from memory would be quite expensive. Instead of time multiplexing, CUDA adopts a streaming approach to virtualization, where thread blocks stream into the machine as long as resources are available, and once full new blocks can only enter the machine as others are retired. This virtualizes the physical number of processors without the performance loss of preemptive time multiplexing, at the cost of restricting the programming model somewhat by requiring that thread blocks be independent.

7.4.2 Coordinating Tasks in Kernels

Once scheduled on an SM, each thread block runs to completion without preemption. This means that thread blocks of a grid cannot safely communicate with each other. A block must never stall waiting for another block to send it some data or signal. The scheduling of thread blocks makes no guarantee on ordering or progress, and because thread blocks are not preemptible, the act of waiting for data in a receiving block may prevent the sending block from ever being scheduled.

The only safe way for separate blocks to freely communicate data among themselves is to write all data to global memory, perform a global synchronization, and read the desired data back. In the CUDA programming model, global synchronization is performed by completing the current kernel and launching a second subsequent kernel. Since all blocks of the first kernel will have completed before any of the second begin, it is safe for any block of the second to read data written by any block of the first. This style of communication is similar to the *bulk synchronous* parallel (BSP) model proposed by Valiant [14] where kernels correspond to the supersteps of a BSP computation.

While blocks of the same kernel may not *communicate* with each other, they may *coordinate* their activities through shared data structures. CUDA defines a number of atomic operations on memory that can be used for this purpose. In CUDA C, they are exposed via intrinsics such as `atomicInc()` to atomically increment the value at an address and `atomicCAS()` to perform an atomic compare-and-swap.

As a simple example, consider a case where each of the p blocks of a kernel computes and stores some per-block partial result, and where the last block to commit its result to memory will read back all partial results and compute a final result. We need a way of determining whether an individual block is the last one to reach a given point in its execution. The code in Figure 7.4 shows a strategy for doing this, using a global counter and atomic memory operations. It first waits for all threads of the block to reach this point and to commit any outstanding memory operations—the use of `__threadfence` here is explained in Section 7.4.3. Next a single thread of the block atomically increments a global counter whose initial value is known to be 0. Since every block will increment the counter by 1 and we know the total number of blocks, we can determine whether this was the last block to increment the counter. Finally, the result of that test is returned to all the calling threads.

The code in Figure 7.4 is actually rather similar to a simplistic implementation of a global barrier, but differs from a barrier in one key respect. No block ever waits for a particular value of the counter to appear; it merely looks at the value and proceeds.

```
__device__ volatile int counter = 0;

__device__ bool am_last_block()
{
    if( gridDim.x==1 )
        // This grid contains only 1 block
        return true;
    else
    {
        __shared__ bool amLast;

        // (1) Make sure all outstanding memory operations complete
        __threadfence();
        __syncthreads();

        // (2) Punch ticket counter and determine if we're last
        if( threadIdx.x==0 )
        {
            amLast = (gridDim.x-1) == atomicInc(&counter, gridDim.x);
            if( amLast )  counter=0;
        }

        // (3) Entire block must wait for result
        __syncthreads();
        return amLast;
    }
}
```

FIGURE 7.4:   A simple procedure to determine whether the calling block is the last block to have reached this point.

It is also important to note that this simple example implementation makes several assumptions that keep it from being useful in all situations. First, its use of a global counter implicitly assumes that only one kernel is running at any one time. Second, it assumes that this grid and its blocks are both one-dimensional. Third, it assumes that it will only be called once per kernel.

Atomic memory operations can also be used to implement many other more complicated shared data structures, such as shared queues or search trees. However, it is vital that, as with our simple counter example, no block ever waits for another block to insert something into the shared data structure. It might be tempting to imagine implementing producer-consumer queues shared by multiple blocks, but this is dangerously susceptible to deadlock if any block ever waits (e.g., rather than exits) when the queue is empty.

7.4.3   Memory Consistency

Our example from the last section illustrates a scenario where data written by one thread block is intended to become visible to one or more other thread blocks in the same grid. In our implementation, we have used a ___threadfence() intrinsic. This intrinsic, which provides a so-called *memory fence*, blocks the calling thread until all its outstanding memory requests have actually been performed. In particular, this means that after the fence, we know that all outstanding writes to global memory performed by this thread have become visible to all other threads. Note that this does not necessarily mean that the written data is physically in memory, however, since other threads may have subsequently written values to the same location.

The memory fence in our example is necessary because we must make sure that all outstanding writes performed by this thread block have been committed before incrementing the counter. The essential role of this counter is to count the number of blocks whose results have been written to memory; hence we must make sure that those results really have been written.

Including such explicit memory fences is necessary because CUDA uses a *relaxed consistency* model [1] for the ordering of operations on external memory. Suppose a thread performs a sequence of writes to global memory, as in *x=a; *y=b;. From the perspective of the writing thread, the value of location x is always written before the value of location y. However, other threads may potentially observe these locations being written in a different order. This can happen for a variety of reasons. Optimizing compilers are allowed to reorder statements if doing so violates no dependencies or other restrictions. Furthermore, the memory subsystems managing the physical DRAM storage attached to parallel processors such as the GPU are often given the freedom to reorder memory requests to different addresses in an effort to improve DRAM performance.

7.5 Writing a CUDA Program

As with any parallel programming paradigm, the CUDA programmer's most important tasks involve formulating a program structure that allows efficient parallel execution. When programming a single parallel device, there are three basic issues that the programmer must address. First, it is obviously necessary to identify the parallel steps of the program that may be implemented as CUDA kernels. Second, the programmer must decompose the computation of a given kernel into parallel tasks that may be assigned to individual thread blocks and determine how the activities of these blocks will be coordinated. Finally, they must select suitable parallel algorithms to accomplish the tasks assigned to individual thread blocks.

To illustrate this process, we will walk through an example of building a CUDA program that performs parallel prefix and reduction operations.

7.5.1 Block-Level Parallel Prefix

In many cases, algorithms that have long been familiar techniques for programming array computers of the past are well-suited to the job of building block-level primitives in CUDA. For example, we will take the classic parallel prefix (or scan)

```
template<typename OP, typename T>
T scan_sequential(OP op, T *values, int size)
{
    for(int i=1; i<size; ++i)
        values[i] = op(values[i-1], values[i]);

    return values[size-1];
}
```

FIGURE 7.5:   Sequential procedure for scan with a generic operator op.

problem [2]. Given a sequence A and a binary associative operator $\oplus$, we want to compute a new sequence B such that $B_0 = A_0$ and $B_i = B_{i-1} \oplus A_i$.

Figure 7.5 shows a direct sequential C++ implementation of a scan procedure. This code is templated so that is generic over the value type T, and performs the scan in place within a single values array. It also expects to be provided with a suitable operator op, which may be defined as an instance of a class such as this:

```
struct plus {
    template<typename T>
    __host__ __device__ T operator()(T a, T b)
    {
        return a+b;
    }
};
```

or through the lambda facility defined in the upcoming C++ standard.

At first glance, the sequential dependency apparent in this loop might suggest that this algorithm is inherently sequential. If we knew nothing about the implementation of op, then this would be true. Our assumption that op is an associative function, free of side effects, allows us to parallelize this computation. Figure 7.6 shows a CUDA C++ implementation of a particularly simple technique for parallel scan. For simplicity, we assume that thread blocks are always 1-dimensional and that there is at least one thread for each input value. The scan_block procedure computes the parallel prefix of the values array in place. Each thread also receives the cumulative value of all input elements.

The algorithm we have used here is well-known and has a long history in the literature. This basic technique has been used for programming parallel prefix operations on a wide range of parallel machines, including SIMD array processors [3,6]. While we include this implementation here because it is quite simple, it is not the most efficient technique. It is theoretically inefficient because it evaluates op(a, b) a total of $O(n \log n)$ times, rather than the optimal $O(n)$ times in the sequential case. More efficient implementations—both in theory and in practice—are described elsewhere [5,12,13] and implementations are available several open source libraries [4,7].

This code also illustrates a common consequence of CUDA's barrier semantics. Notice that the body of the loop takes the following form:

```
template<typename OP, typename T>
__device__ T scan_block(OP op, T *values,
                        int size, int i=threadIdx.x)
{
    for(int h=1; h<size; h *= 2)
    {
        bool active = (i>=h && i<size);
        T partial;

        // Grab partial result from thread i−h where h = 2^d
        if(active)  partial = values[i-h];
        __syncthreads();

        // Update partial result for thread i
        if(active)  values[i] = op(partial, values[i]);
        __syncthreads();
    }

    // Every thread returns cumulative value
    return values[size-1];
}
```

FIGURE 7.6: Procedure to perform parallel prefix (or scan) with a generic operator op within a thread block.

```
if(P) { ... }
__syncthreads();
if(P) { ... }
__syncthreads();
```

It might at first seem that this code could be written more compactly by removing the duplicate if(P) statements and placing everything inside the body of a single conditional. However, this would violate the requirement that barriers can only exist inside conditionals when all threads of the block will evaluate the conditional in the same way. Since the active predicate depends on the thread index, it will be evaluated differently by different threads. The condition of the loop, on the other hand, is evaluated identically by every thread and can therefore safely contain barriers within its body.

7.5.2 Array Reduction

Details on building complete scan kernels can be found in one of the several available papers on CUDA scan techniques [5,12,13]. For our purposes here, we will illustrate the coordination of multiple blocks within a single kernel by using this scan primitive to implement an array *reduction* procedure. Reduction is a simpler operation than scan. It takes a sequence $A = (a_0, \ldots, a_{n-1})$, a binary commutative function $\oplus$, and computes the value of $a_0 \oplus \cdots \oplus a_{n-1}$. Common special cases of reduction,

```
template<typename OP, typename T>
__device__ T reduce_array(OP op, T identity,
                          const T *begin, const T *end,
                          T *scratch,
                          int i=threadIdx.x,
                          int blocksize=blockDim.x)
{
    T sum = identity;
    int width = min(int(end-begin), blocksize);

    // Grab values from array one block-sized tile at a time,
    // accumulating them into sum as we go.
    while(begin + i < end )
    {
        sum = op(sum, begin[i]);
        begin += blocksize;
    }

    // Place partial sums in scratch space
    scratch[i] = sum;
    __syncthreads();

    // Perform parallel reduction of per-thread partial results
    return scan_block(op, scratch, width, i);
}
```

FIGURE 7.7:   Parallel reduction of an array using a single thread block.

such as sum, are provided as built-in array primitives in systems like Fortran 90 and as special collective operations by parallel programming frameworks like OpenMP and MPI.

Figure 7.7 demonstrates how to reduce an array of arbitrary size with a single thread block. It accepts a generic operator op and a corresponding identity value. It expects pointers to the beginning and end of the input array, a convention we adopt following the C++ Standard Template Library, and a pointer to a "scratch" space. The input array bounded by begin and end can live in either global or shared on-chip memory and be arbitrarily long. The scratch array should contain space for one value per thread and, while not required for correctness, should be in shared on-chip memory for better performance.

The first part of this procedure is a loop during which the block iterates over block-sized tiles of the input array. In the first iteration, the blocksize threads of the block will load elements 0 through blocksize-1 with each thread accumulating its corresponding value into the sum variable that holds its running total. Each thread offsets its reading location by blocksize and repeats. This is essentially the pattern of *strip mining* commonly used on vector machines when looping over arrays. This access pattern guarantees that contiguously numbered threads i and $i + 1$ always access contiguous memory locations in the input array. In turn, this allows the GPU's hardware to *coalesce* these adjacent loads into a minimal number of transactions with external memory.

After each thread has accumulated a partial result for its slice of the input array, every thread writes its partial result into the scratch space. They then collectively execute the `scan_block` procedure that we described earlier. The return value of this function is the final combination of all the partial results which it was given, which is precisely the result of the reduction we are seeking to compute. This performs a bit more work than necessary, since we require only a reduction rather than a full prefix sum. We use `scan_block` simply to avoid the need to give block-level implementations of both scan and reduce routines.

7.5.3 Coordinating Whole Grids

Having built the code to reduce an array of values with a single thread block, we now consider how to coordinate multiple blocks in order to reduce large arrays with an entire grid. We will focus on the specific case of summing arrays of integers. The structure of reductions lends itself to a particularly simple division of labor between thread blocks. Given n input values, we select some number p of thread blocks to use; the kernels we present work for any choice of p. We can then split the input sequence into p contiguous blocks and launch a grid using a single thread block to reduce each subsequence to a single value. Launching a second grid with 1 thread block can reduce the p partial sums to the final sum. This pattern is shown in the code in Figure 7.8. It accepts an input array (`values`) and uses the `results` array to hold partial sums. When the `sum()` procedure has terminated, the total sum of `values` can be found in `results[0]` which resides in device memory.

The number of thread blocks p to use in this example can be determined in a variety of ways, and the optimal choice will generally depend on the specific hardware being used. Typical choices are to divide n by the target block size, to use a fixed size p that is related to the physical number of multiprocessors, or some other related heuristic. In our example, we use a value of $p = \min(n/1024, 512)$. The aim of this heuristic is to make sure each block has some reasonable amount of work to do—in this case 8 elements per thread of a 128-thread block—while keeping the number of blocks modest (512) so that the single block of the second kernel will not have too many partial sums to combine. We make no claim of optimality for this specific heuristic, but it does illustrate the frequent trade-off involved in balancing the amount of parallelism created versus the amount of work available for each block to process.

As we have already seen, the NVIDIA CUDA C compiler (`nvcc`) supports C++ templates even though it does not currently provide support for the complete C++ language. This support for templates, and in particular the compile-time instantiation of templated kernels, opens up the possibility of packaging CUDA kernels within higher-level C++ template libraries. One example of this programming approach is Thrust [7], an open-source library that provides an STL-like collection of templated containers and algorithms that are coupled together with iterators. Figure 7.9 shows a simple program using the `thrust::reduce` algorithm to sum all the elements in a

```
__global__ void sum_kernel(const int *values, int size, int *results)
{
    extern __shared__ int scratch[];

    int per_block = (size + gridDim.x - 1) / gridDim.x;
    const int *begin = values + blockIdx.x * per_block;

    const int *end = begin+per_block;
    if( end>values+size )   end=values+size;

    int sum = reduce_array(plus(), 0, begin, end, scratch);

    if( threadIdx.x==0 )
        results[blockIdx.x] = sum;
}

__host__ void sum(const int *values, int size, int *results)
{
    // Use 128-thread blocks
    int B = 128;

    // Give at least 1024 elements to each block, but
    // don't use more than 512 blocks.
    int P = min(size/1024, 512);

    // Each block needs a B-element scratch space
    int nbytes = B*sizeof(int);

    // Produce partial sums for P blocks ...
    sum_kernel<<<P, B, nbytes>>>(values, size, results);
    // ... and combine P partial sums together into final sum
    sum_kernel<<<1, B, nbytes>>>(results, P, results);
}
```

FIGURE 7.8:   Code for summing entire arrays using parallel reduction spread across an entire grid of thread blocks.

vector of randomly generated integers. The vector containers provide an interface like that of the STL `std::vector` container, while transparently managing movement of data between host and device memory.

7.6   GPU Architecture

Readers may find it helpful in optimizing CUDA programs to understand additional details regarding the hardware organization. NVIDIA introduced GPU computing with its first CUDA-capable architecture that was first released in late 2006. In 2009,

```
int main()
{
    // Generate random data on the host
    thrust::host_vector<int> x(1000000);
    thrust::generate(x.begin(), x.end(), rand);

    // Transfer to device and reduce
    thrust::device_vector<int> dx = x;
    int sum = thrust::reduce(dx.begin(), dx.end(), plus());

    // Print result and exit
    printf("Sum=%d\n", sum);
    return 0;
}
```

FIGURE 7.9: Reduction example using the Thrust template library.

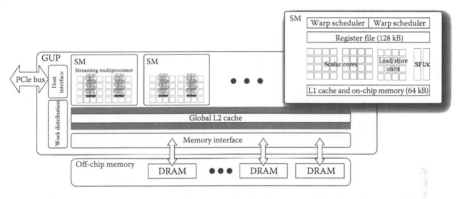

FIGURE 7.10: High-level architectural block diagram of a Fermi-class GPU.

NVIDIA introduced a radically new architecture code-named "Fermi" (Figure 7.10). In this section, we describe some of the most important properties of this generation of GPUs.

In NVIDIA GPUs, the instruction set exposed to the user-level software—PTX—is a RISC instruction set that is mapped onto the native machine code executed by the GPU. This provides greater flexibility to change hardware details as technology changes. It also improves portability, because PTX is intended for efficient binary-to-binary translation, with no need for an interpreter. PTX can also provide illusion of an infinite register file to higher-level compilers, since register allocation is performed during translation to the native ISA.

For most purposes, it is sufficient to assume that every thread executes on its own virtual processor. From a performance optimization standpoint, it is sometimes useful to know that in reality, the many threads and associated virtual processors of a thread block are multiplexed onto a smaller number of physical processing elements. Each of these scalar processing elements is deeply multithreaded; each of Fermi's processing elements have 48 concurrent thread contexts. These threads are time-multiplexed onto the hardware. When one thread stalls for any reason, other threads that are ready to execute may proceed. The hardware imposes zero latency for switching among threads, so a different thread can run every cycle. This means that long-latency operations, such as references to off-chip global memory, do not necessarily reduce aggregate processing throughput, as long as other threads are ready to execute.

Each thread block operates on one SM, so that all the block's threads can share an L1 cache and a block of on-chip shared memory. Thread blocks are arranged in *warps* of 32 threads that operate in lockstep, often referred to as "single instruction, multiple data" or SIMD operation. Note that with each thread having its own program counter and scalar register file, this is an array style of SIMD that differs from the more familiar vector style of SIMD, a distinction NVIDIA captures with the term SIMT for "single instruction, multiple thread." The thread scheduler for a thread block actually schedules warps rather than individual threads. There is no need for different warps from the same thread block to operate in lockstep. In fact, there is no need for threads in the same warp to operate in lockstep; unlike vector architectures, branching (including indirect branching) is fully supported in hardware. However, since the underlying hardware units executing the warp are lockstepped and must follow the same execution sequence at any one time, any threads within the same warp that are following different execution paths are masked off, and the warp is executed again with a different mask for each unique execution path. For compute-bound applications, this will prevent reaching peak throughput. In short, this SIMT hardware organization only affects performance when threads in the same warp branch in different directions, and managing divergence is a performance optimization instead of a correctness concern. Fortunately, divergent branching can often be eliminated with minor algorithmic modifications and careful data layout.

SIMT improves area efficiency by allowing the processing elements to share fetch, decode, issue, and scheduling hardware. The SM pipeline actually operates at the granularity of warps and consists of fetch, decode, issue, scheduling, dispatch, execute, and writeback stages. The issue stage is responsible for tracking and managing instruction dependencies to determine, for each warp, when its next instruction may execute. The scheduling stage chooses which of the warps with ready instructions is actually allowed to move on to execution. Dispatch is responsible for fetching operands and sending the instruction with its operands to the execution units and these control signals are fanned out to all hardware units. If thread blocks do not require too much space in the register file and shared memory, an SM can support two or more thread blocks, but the scheduler treats all warps the same for scheduling purposes. The distinction between thread blocks chiefly matters in keeping their shared-memory allocations separate.

In Fermi, each SM contains 32 floating-point units, 32 integer units, 16 load-store units, and 4 "special functional units" that support complex functions such as transcendentals, and each SM can issue instructions from two warps per cycle. Fermi supports the new IEEE 754-2008 floating-point standard for both single- and double-precision. A notable addition over the 754-1985 standard is support for fused multiply add (FMA), which combines a multiply and add, using a single rounding step to avoid any loss of precision. For Fermi-generation Tesla products, floating-point double-precision throughput is one half of single-precision throughput, giving 515 GFLOP/s peak throughput for double precision and 1.03 TFLOP/s for single precision. Tesla products also provide ECC protection on all memory structures, including the register file, caches and shared memory, and external DRAM. The GDDR5-based memory interface memory interface delivers very high bandwidth to external memory—for example, the Tesla C2050 provides 144 GB/s peak bandwidth.

One of the most important architectural differences in Fermi compared to other GPU architectures is that it adds a cache hierarchy to the global address space, with 64 kB first-level capacity per SM and a large (768 kB) global L2. The L1 is divided between cache and per-block shared memory, with either 48 kB of L1 and the conventional 16 kB shared memory, or vice versa. It is important to note that the cache hierarchy does not support hardware cache coherence. If coherence is required, it must be supported in software by flushing writes to L2.

All CUDA-capable architectures also support cached texture memory. Texture memory has been a fixture of 3D graphics hardware for many generations, because it provides high-bandwidth support to multiple, neighboring cells within an array. This makes it useful in general-purpose computing for some kinds of read-only array accesses. Even given the data caches provided by Fermi, using these texture caches can provide performance benefits since using them increases the aggregate on-chip cache capacity available to the program.

Another important property of the memory system is that accesses to global memory are wide, possibly as wide as a warp, and *coalesced*. Full bandwidth is achieved when all threads in a warp load from a single cache line. This allows the loads from all threads in a warp to be serviced in a single transaction to the global memory. This preferred "SIMD-major" order differs somewhat from the situation on multicore CPUs which typically prefer that individual threads access contiguous memory locations themselves.

Unlike conventional CPUs, atomic read-modify-write operations do not fetch the memory location into the SM, as conventional CPUs do. Instead, the atomic memory operation is sent to the memory subsystem and waits there until it can complete. This means that SM compute resources are not tied up waiting for completion, although the specific warp that issued the atomic may be delayed. Fermi dramatically improves the bandwidth of atomic and memory-fence operations compared to prior generations.

Among all these aspects of the memory hierarchy, the ones that are most important will vary from application to application, and it is always important to identify the specific bottleneck. In general, for applications that are memory bound, minimizing contention for the memory channels will be most important. This can be achieved by effective use of the per block shared memory and the caches, thus reducing memory bandwidth utilization; and by maximizing coalescing when memory is accessed. Maximizing the number of concurrent threads is often helpful too, because this increases latency tolerance; but only when increasing the thread count does not increase the cache miss rate. Optimizing memory access patterns will often be more important than execution divergence within a warp.

7.7　Further Reading

In this chapter, we have provided only a quick overview of the CUDA architecture for parallel programming and the GPU hardware on which it runs. There are a number of other sources available for further information on CUDA, GPU hardware architecture, and efficient parallel programming techniques for these systems.

The CUDA Toolkit, which includes the NVIDIA CUDA C compiler and related development tools, contains both the CUDA Reference Manual [11] and Programming Guide [10], which document the CUDA APIs and C language extensions. A companion manual, the CUDA Best Practices Guide [9], outlines numerous techniques for improving the efficiency and reliability of CUDA C code. NVIDIA's GPU Computing SDK includes several programming examples, including one that examines the efficient implementation of reduction kernels. Finally, NVIDIA's CUDA Zone—online at http://www.nvidia.com/CUDA—contains a directory of numerous third-party projects and papers using CUDA to solve real-world problems.

Finally, *Programming Massively Parallel Processors: A Hands-On Approach*, by Kirk and Hwu [8], provides many useful examples to illustrate effective parallel programming for GPUs.

References

1. S. V. Adve and M. D. Hill. A unified formalization of four shared-memory models. *IEEE Trans. Parallel Distrib. Syst.*, 4(6):613–624, 1993.

2. G. E. Blelloch. *Vector Models for Data-Parallel Computing*. MIT Press, Cambridge, MA, 1990.

3. W. J. Bouknight, S. A. Denenberg, D. E. McIntyre, J. M. Randall, A. H. Sameh, and D. L. Slotnick. The Illiac IV system. *Proc. IEEE*, 60(4):369–388, April 1972.

4. CUDPP: CUDA data-parallel primitives library. http://www.gpgpu.org/developer/cudpp/, July, 2009.

5. Y. Dotsenko, N. K. Govindaraju, P.-P. Sloan, C. Boyd, and J. Manferdelli. Fast scan algorithms on graphics processors. In *Proceedings of the 22nd Annual International Conference on Supercomputing*, Island of Kos, Greece, pp. 205–213. ACM, New York, June 2008.

6. W. Daniel Hillis and G. L. Steele, Jr. Data parallel algorithms. *Commun. ACM*, 29(12):1170–1183, 1986.

7. J. Hoberock and N. Bell. Thrust: A parallel template library. http://www.meganewtons.com/, March 2010. Version 1.2.

8. D. Kirk and W. Hwu. *Programming Massively Parallel Processors: A Hands-On Approach*. Morgan Kaufmann, San Francisco, CA, 2010.

9. NVIDIA Corporation. *NVIDIA CUDA C Programming Best Practices Guide*, July 2009. Version 2.3.

10. NVIDIA Corporation. *NVIDIA CUDA Programming Guide*, July 2009. Version 2.3.

11. NVIDIA Corporation. *NVIDIA CUDA Reference Manual*, July 2009. Version 2.3.

12. S. Sengupta, M. Harris, and M. Garland. Efficient parallel scan algorithms for GPUs. Technical Report NVR-2008-003, NVIDIA, December 2008.

13. S. Sengupta, M. Harris, Y. Zhang, and J. D. Owens. Scan primitives for GPU computing. In *Graphics Hardware 2007*, San Diego, CA, pp. 97–106. ACM, August 2007.

14. L. G. Valiant. A bridging model for parallel computation. *Commun. ACM*, 33(8):103–111, 1990.

Chapter 8

Programming the Cell Processor

Christoph W. Kessler

Contents

8.1 Introduction

Cell Broadband Engine™,* often also just called "the Cell processor," "Cell/B.E." or shortly "Cell," is a heterogeneous multicore processor architecture that was introduced in 2005 by STI, a cooperation of Sony Computer Entertainment, Toshiba, and IBM. With its special architectural design, Cell can speed up gaming, graphics, and scientific computations by up to two orders of magnitude over contemporary standard general-purpose processors, and this at a comparable power consumption. For instance, in a comparative simulation experiment [58], a Cell processor clocked at 3.2 GHz achieved for a single-precision dense matrix-matrix multiplication (SGEMM) a performance of 204.7 Gflops† at approximately 40 W, while (at that time) an AMD Opteron™ processor only reached 7.8 Gflops and an Intel Itanium2™ only 3.0 Gflops on this benchmark.

Most prominently, Cell is currently used as the computational core in the Sony PlayStation® 3 (PS3), a high-performance game console that was released by Sony in 2006 and has, by September 2010, been sold in more than 47 million units worldwide [52], but also in the fastest supercomputer of the world 2008–2009, *RoadRunner* at Los Alamos National Laboratory in New Mexico, USA, which combines the performance of 12,960 PowerXCell™ 8i processors and 6,480 AMD Opteron™ dual-core processor to a peak performance of 1.026 Petaflops. Details about the architecture of Cell will be given in Section 8.2.

A drawback is that Cell is quite hard to program efficiently. In fact, it requires quite some effort in writing code that comes halfway close to the theoretical peak performance of Cell. In Section 8.3 we will present some of the techniques that are necessary to produce code that utilizes the hardware resources efficiently, such as multithreading, SIMD computing, and multiple buffering of DMA communication. Our example codes in C are based on the Cell software development kit (SDK) from IBM [28], which we briefly summarize in Section 8.4.

The programmability issue has spawned various efforts in both academic research and software industry for developing tools and frameworks for more convenient programming of Cell. We report on some of these in Section 8.5.

A selection of algorithms tailored for Cell is listed in Section 8.6.

Note that a single chapter like this cannot replace a comprehensive tutorial to Cell programming with all necessary technical details; we can here just scratch the surface. We therefore focus on the most important system parts and programming

---

* A list of trademarks mentioned in this chapter is given at the end of this chapter.
† IBM confirmed 201 Gflops for SGEMM by measurements [58].

techniques for high-performance programming on Cell, and refer the reader to the relevant literature for further details.

8.2 Cell Processor Architecture Overview

Figure 8.1 gives an overview of the main architecture components in the Cell processor.

Cell has nine processing elements of two different kinds: one *PowerPC processor element* (PPE) and eight *synergistic processor elements* (SPEs).

On earlier versions of the PS3, where it is possible to install an alternative operating system such as Linux, six SPEs are accessible to the application programmer at the Linux level. Sony removed this alternative OS feature in the most recent PS3 (Slim) edition.

8.2.1 Power Processing Element

The PPE contains one general-purpose processor core, called the *power processor unit* (PPU), which is a standard 64-bit PowerPC core with Vector/SIMD Multimedia extension (VMX); and the *power processor storage subsystem* (PPSS) that handles memory requests to and from the PPE and controls the level-2 (L2) cache in the PPE, which is 512 kB large.

The PPE acts as the master processor of Cell. In principle, it could be used just as an ordinary PowerPC processor by simply ignoring the remaining elements. For instance, binary programs written for the PowerPC 970 processor can execute on Cell without modification [24]. Beyond the standard PowerPC instruction set, the PPU also supports the AltiVec instruction set and additional 128-bit vector data types for SIMD programming. Accordingly, beyond its 32 64-bit general-purpose registers and 32 64-bit general-purpose floating-point registers, the PPU also has 32 vector

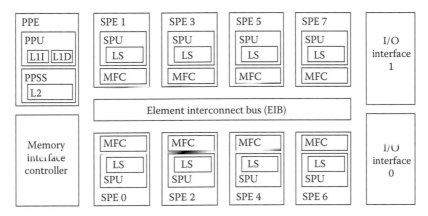

FIGURE 8.1: Cell processor overview.

registers of 128 bit size. The PPU has a level-1 (L1) instruction cache and L1 data cache of 32 KB each.

The PPU supports two hardware threads, which look to the operating system like two independent processors in a multiprocessor with shared memory. While each thread has its own instance of the registers, the program counter, and further resources holding its state, most other resources are shared.

Only the PPE runs a full-fledged operating system. For SDK-level Cell programming, this is usually a version of Linux. For instance, the IBM SDK 3.1 [28] is available for Red Hat Enterprise Linux 5.2 and Fedora 9.

Virtual memory is organized in segments and these in turn in pages. The PPU has a memory management unit (MMU) that translates virtual to physical memory addresses. A translation look-aside buffer (TLB) caches page table entries to speed up translation between virtual and physical memory addresses.

All values and instructions are stored in big-endian way.

8.2.2   Synergistic Processing Element

The SPEs are programmable slave processor cores. Each SPE mainly consists of a so-called *synergistic processor unit* (SPU), which is a special-design 128-bit RISC processor with a small local memory, and of a communication coprocessor (MFC) described later.

In contrast to the PPU, the SPU has no caches. The SPU's load and store instructions directly access its local on-chip RAM of 256 kB size, called the *local store* (LS), which has to accommodate both code and data of SPE programs that are being accessed by the SPU. The programmer is responsible for moving code and data explicitly from main memory to the local store, using DMA operations that will be described later. SPE programs are often data-intensive and therefore can only keep a small part of their total data in the local store at a time; the remainder is to be kept in off-chip main memory until it is used. Also, longer SPE programs need to be split into smaller parts and these parts be loaded into the local store as needed, in order to fit in its very small size.

The fact that the local store is not a hardware-controlled cache but under full control of the application program allows for more predictable performance but puts a larger burden on the programmer. Many high-level programming models for Cell (see Section 8.5) attempt to abstract from the memory structure and explicit data transfer. For instance, a software cache solution for the local store has been provided by IBM in their OpenMP compiler [14].

The SPU is a relatively simple RISC processor architecture that was especially designed for accelerating SIMD computations.* The SPU has 128 registers of 128 bit

---

* SIMD (single instruction stream, multiple data streams) computing is a special control paradigm of parallel computing where the same instruction is concurrently applied to multiple elements of an operand vector; see Section 8.3.4 for further explanation. Most desktop and server processors feature SIMD instructions in some form today. Cell implements SIMD instructions in both PPU and SPU. SIMD parallelism should not be confused with the more general concept *data parallelism* which also allows more complex operations, not just individual instructions, to be applied element-wise.

size each. All arithmetic SPU instructions operate on 128-bit words in 128-bit SPU registers; SPU load instructions move 128-bit words from the SPE's local store to the SPU registers, and store instructions work vice versa. A 128-bit register can contain so-called *vector datatypes* that consist of either two adjacent 64-bit double-precision floating-point or long-integer words, or four 32-bit single-precision floating-point or integer words, or eight 16-bit short integers, or 16 eight-bit characters, where the integral data types are available in both signed and unsigned form (see also Figure 8.8). In addition, *scalar data types* are also supported, but these occupy only a part of a 128-bit register and leave the rest unused (see also Figure 8.7). A computation on, for instance, a single 32-bit single-precision floating-point word is actually more expensive than the same operation applied to a vector of four 32-bit floating-point words stored consecutively in the same SPU register: That one word may first need to be extracted by bitwise mask operations from its current position in its 128-bit register, then, for binary operations, possibly shifted to obtain the same relative position in a 128-bit word as the other operand, and finally the result needs to be shifted and/or inserted by bitwise masked operations into its final position in the destination register.

It should also be noted that the SPU implementation of floating-point arithmetics deviates from the IEEE-754 standard in a few minor points, see, for example, Scarpino [50, Ch. 10.2].

The SPU also has certain support for instruction-level parallelism: The seven functional units of the SPU are organized in two fully pipelined execution pipelines, such that up to two instructions can be issued per clock cycle. The first pipeline (called "even") is for instructions of the kind arithmetic, logical, compare, select, byte sum/difference/average, shift/rotate, integer multiply-accumulate or floating-point instructions; the second pipeline (called "odd") is for instructions such as shift/rotate, shuffle, load, store, MFC channel control, or branch instructions. Instruction issue is in-order, which (in contrast to current out-of-order issue superscalar processors) relies more on the compiler's instruction scheduler to optimize for a high issue rate by alternating instructions of "odd" and "even" kind. The latency for most arithmetic and logical instructions is only 2 clock cycles, for integer multiply-accumulate 7 cycles, and for load and store 6 cycles. Branches on the SPU can be quite costly; depending on whether a branch is taken or not taken, its latency is between 1 and 18 clock cycles. There is no hardware support for automatic branch prediction. The compiler or assembly-level programmer can predict branches and insert branch hinting instructions in an attempt to minimize the average case latency. Fortunately, the high-level programmer is hardly concerned with these low-level details of instruction scheduling, but when tuning performance it may still be useful to be aware of these issues.

The SPEs have only a very simple operating system that takes care of basic functionality such as thread scheduling. All I/O has to be done via the PPE. For instance, certain I/O functionalities such as `printf` that are supported in the C library of the SPEs call the PPE for handling.

In the early versions of Cell, the SPEs' double precision performance was far below that of single precision computations. In 2008, IBM released an updated Cell version called *PowerXCell 8i*, where the accumulated SPE double-precision peak performance has improved from 14 to 102 Gflops. This Cell variant is used, for example, in the most recent version QS22 of IBMs dual-Cell blade server series.

Main differences between PPE and SPE: While the general-purpose PPE is intended to run the top-level program control, I/O, and other code that cannot effectively utilize the SPE architecture, the SPEs are generally faster and more power-efficient on vectorizable and/or parallelizable computation-intensive code. Hence, the SPEs are intended to work as programmable accelerators that take over the computationally heavy phases of a computation to off-load the PPE.

PPU and SPU have different address sizes (64 and 32 bit, respectively) and different instruction sets and binary formats; so PPU and SPU codes each need to be produced with their own processor-specific toolchain (compiler, assembler, linker).* Also the APIs for DMA memory transfers and for vector/SIMD operations differ (slightly) for PPU and SPU. This heterogeneity within the system additionally contributes to the complexity of Cell programming.

8.2.3   Element Interconnect Bus

The PPE and the SPEs are connected to each other, to the I/O interface and to off-chip main memory via the Element Interconnect Bus (EIB). The EIB operates at half the processor clock frequency, that is, 1.6 GHz. Overall, there are 12 bus units (access links): the PPE, the eight SPEs, the interface to off-chip main memory, and two I/O interfaces. Each bus unit has a peak bus access capacity of 16 bytes per clock cycle (i.e., 25.6 GB/s), 8 bytes per direction. Note that this also implies that the maximum bandwidth to/from off-chip main memory cannot exceed 25.6 GB/s per direction. In practice, up to 24.6 GB/s can be achieved [9].

The EIB has a theoretical peak capacity of 204.8 GB/s, which is determined by the maximum rate at which the bus units can snoop addresses, namely, one address per bus cycle, where each snooped address request may transfer up to 128 bytes [40].

However, the actual EIB capacity may strongly depend on the concrete communication pattern. For instance, contention occurs if multiple data transfers target the same bus unit simultaneously.

Also, the internal implementation of the EIB has some influence on communication performance: The EIB consists of two separate communication paths: a star network for commands (i.e., transfer requests with source and target address, etc.) and a data network consisting of four parallel unidirectional rings, two per direction; see Figure 8.2 for the relative positions of communicating bus units on the rings. A hardware bus arbiter assigns dynamically available ring segments between bus units to data transfer requests along the shortest path between the two communicating bus units, and thereby provides the bus-like abstraction of the underlying quad-ring network to the programmer. On each ring, up to three data transfers can proceed

---

* A single executable file holding a Cell program can be obtained by embedding the SPU binary code and data segments as a special data segment into the PPU binary file, see Section 8.3.1.

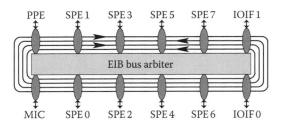

FIGURE 8.2: The four unidirectional rings and bus arbiter of the Element Interconnect Bus (EIB) network of Cell.

concurrently as long as their assigned ring segments do not overlap. Actual EIB bandwidths between 78 and 197 GB/s have been observed in an experiment that varied the positions of four communicating pairs of SPEs [9].

8.2.4 DMA Communication and Memory Access

A SPU can only access its local store. All accesses to the EIB go via the SPE's *Memory Flow Controller* (MFC), which is a communication coprocessor that offloads the SPU from data transfer tasks.

Data transfers are realized via *Direct Memory Access* (DMA). The DMA transfer API on the SPE side contains, in particular, the functions `mfc_get` and `mfc_put` which issue DMA requests by the calling SPE to move data to and from this SPE, respectively. The parameters of a DMA request contain, in particular, the source address, the destination address, and the size of a block of bytes to be transferred. Source addresses (for `mfc_get`) and destination addresses (for `mfc_put`) refer to external memory (main memory or some other SPE's local store) and are therefore to be specified as *effective addresses*, i.e., system-wide unique addresses that are mapped to the virtual address space of the PPE process, where all SPE local store address spaces and the main memory address space are embedded. Local store addresses are represented by 32-bit words. Effective addresses are either 64-or 32-bit wide, depending on whether the PPE process is configured as a 64-bit or 32-bit process. Clearly, PPE code and SPE code of the same application must use the same width for effective addresses. Virtual addresses are 65 bit wide.

DMA block sizes can be as small as 1, 2, 4, 8 or 16 bytes, or any 16-byte multiple up to 16384 bytes. Blocks that exceed 16,384 bytes in size must be split in several blocks of up to 16,384 bytes size, and these must be handled separately in several DMA transfers. For transfer of data of less than 16 bytes size, addresses must be naturally aligned, the four least significant bits of source and destination address must be equal; addresses of transfers of 16 bytes or larger must be 16-byte aligned. DMA transfer is most efficient if blocks are at least 128 bytes large, preferably a multiple of 128 bytes in size, and both source and destination address are 128-byte aligned. In general, DMA transfer is more efficient (per transferred byte) for larger blocks.

The mapping from process-independent, physical local store addresses and effective addresses (used on the SPE side) to virtual memory addresses of the executing process (provided by the operating system) and vice versa is done by a memory management unit (MMU) in each MFC. Also, the MMU takes care of access permission control. For these mappings, the SPE MMUs use the same translation mechanism as the PPE MMU, that is, with page and segment tables as in the PowerPC architecture. In particular, the 36-bit segment IDs in effective addresses are translated to the 37-bit segment IDs in virtual addresses, using a segment look aside buffer in memory.

The PPE can query the effective addresses, for example, of the local store of an SPE on which it has created an SPE context, using special API functions such as `spe_ls_area_get`, and pass them to SPE threads, for example, in order to enable SPE-to-SPE DMA communication. In principle, the PPE could then even directly access these addresses, as they have been mapped into its virtual address space; however, DMA transfers to and from local store are more efficient.

The DMA controller in the MFC executes DMA commands in parallel to the SPU. DMA transfers work *asynchronously* to the SPU computation: The SPU issues a request to the MFC and continues with the next instructions; it may then poll the status of the DMA transfer, which can be necessary to resynchronize with the issuing SPU thread.*

In order to keep track of all pending DMA requests issued by the SPU, each one is given a *tag*, an integer between 0 and 31. DMA requests with same tag belong to the same *tag group*; for instance, the multiple requests into which the transfer of a large data block is to be split will all have the same tag and belong to the same tag group. Synchronization of the SPU with pending DMA transfers can be in terms of tag group masks, that is bit vectors where each bit position corresponds to one of the 32 tag groups. In this way, it is possible to selectively wait for completion of any or all from an arbitrary subset of DMA transfers.

The MFC has two queues for buffering DMA requests to be performed: The *MFC SPU command queue* for DMA requests issued by the SPU, and the *MFC proxy command queue* for DMA requests issued by the PPE or other devices to this SPE, which are entered remotely by appropriate load and store instructions to memory-mapped I/O (MMIO) registers in the MFC. Up to 16 issued DMA transfer requests of an SPE can be queued in the MFC SPU command queue. If the MFC command queue is full and the current SPU instruction wants to issue another DMA request, that SPU DMA instruction will block until there is a free place in the queue again. Within a single tag group, the MFC queueing mechanism implements not necessarily a FCFS (first come first served) policy, as requests may "overtake" each other. For this reason, the DMA API provides variants of `mfc_put` and `mfc_get` that enforce a partial ordering of DMA requests in the queue.

Special *atomic DMA routines* are provided for supporting concurrent write accesses to the same main memory location. These use mutex locking internally, where the unit of locking (and thereby atomicity) are (L2-)cache line blocks in memory, that is, 128 bytes that are 128-byte aligned.

---

* Programmers familiar with the Message Passing Interface (MPI) will recognize the similarity to the non-blocking (incomplete) message passing operations in MPI, such as `MPI_Isend`, and corresponding probe and wait operations, such as `MPI_Test` and `MPI_Wait`, respectively.

The synchronization library also provides common mutual exclusion synchronization functionality such as mutex_lock, mutex_trylock and mutex_unlock on mutex locks, that is, signed integer variables in main memory, reader-writer locks, condition variables, and common atomic update functions on individual integer memory words, such as atomic_add, atomic_inc, atomic_dec, etc.

DMA request lists are an abstraction for realizing scatter and gather DMA operations,* that is, reading data that comes from multiple separate source locations in memory into a contiguous block in local store, or vice versa, writing a contiguous block to multiple destination locations that are spread out over main memory. A list of such multiple locations can contain up to 2048 memory addresses.

The *DMA latency* is the time elapsed between starting issuing a DMA command to the MFC and starting writing the transferred data to the target address. The largest components of DMA latency are caused by bus snooping for the (PPE) cache coherence protocol and for the actual data transfer between SPEs or between SPE and main memory, while the overhead of issuing the DMA command to the MFC and processing it in the MFC is relatively small. If all required resources are immediately available, the latency of a blocking DMA transfer of up to 512 bytes (i.e., 4 cache lines) is about 90–100 ns for puts and for gets to local store; for gets from main memory, DMA latency is somewhat higher, about 160 ns [40].

8.2.5 Channels

Channels are a fast message-passing communication mechanism for sending 32-bit messages and commands between different units on Cell. For instance, a SPU communicates with its local MFC via SPU channel instructions, which are accessible to the programmer as functions in a special SPE Channel API, such as spu_readch (ch) (read from channel ch) and spu_writech(ch, intval) (write integer value intval to channel ch).

Channel messages are intended for signaling the occurrence of specific events, for synchronization, for issuing DMA commands to the MFC, for querying MFC command status and command parameters, and for managing and monitoring tag groups. For instance, the DMA operations mfc_put and mfc_get each are implemented as 6 subsequent channel write operations that write the DMA parameters (addresses, size, tag) and the MFC command code itself.

An MFC provides 32 channels. Each channel is for either read or write access by the SPU. A channel is a buffer for channel messages and has a limited *capacity* (number of 32-bit entries), where the number of remaining free entries is given by a counter indicating its *remaining capacity* at any time. Most channels have a capacity of one; the channel for writing DMA commands issued by the SPU has a capacity of 16 entries (see above), and the channel for reading from the SPU's in-bound mailbox (see below) has capacity 4.

---

* These should not be confused with the collective communication operations MPI_Scatter and MPI_Gather in MPI, which involve a group of several processors. Here, scatter and gather apply to a single SPE.

A channel is either *blocking* or *nonblocking*: If the SPE attempts to, for instance, write a blocking channel whose capacity counter is 0 (i.e., the channel is full), the SPE will stall (i.e., switch to low-power state) until at least one pending command in that channel has been executed by the MFC to free an entry, and a corresponding acknowledgment is received from the channel. If this behavior is undesirable, the capacity counter should be checked by `spu_readchcnt( ch )` before writing. For example, the channel for issuing MFC DMA commands is a blocking channel. For a non-blocking channel, writing will always proceed immediately.

Reading from and writing to channels are atomic transactions. Channel operations are performed in program order.

PPU-initiated DMA commands to access an SPE's local store do not use MFC channels but write via the EIB into the MMIO registers in the MFC, which are made globally accessible by mapping them to effective addresses.

8.2.6   Mailboxes

Mailboxes are message queues for indirect communication of 32-bit words between an SPE and other devices: The recipient or sender on the other end of a mailbox communication is not directly addressed or predefined, as opposed to DMA channel communication; for instance, it may be other SPEs or the PPE who fetch a message from an SPE's out-bound mailbox or post one into an SPE's in-bound mailbox.

A SPE has four in-bound mailboxes that the SPU accesses by read-channel instructions, and one out-bound mailbox and one out-bound interrupt mailbox that the SPU accesses by write-channel instructions.

8.2.7   Signals

Signal communication is similar to mailbox communication in that both transfer messages that consist of a single 32-bit word. In contrast to mailbox messages, signals are often used in connection with DMA communication for notification purposes, and especially for direct communication between different SPEs. Also, broadcasting a single message to many recipients is possible with signals.

Signals are implemented on top of MFC channel communication, using the two channels `SPU_RdSigNotify1` and `SPU_RdSigNotify2` dedicated to this service. Signals can be received via these channels by the API functions `spu_read_signal1` and `spu_read_signal2`, respectively. One can probe for expected signals by `spu_stat_signal1` and `spu_stat_signal2`, respectively. These functions do not block but return 1 if a signal is present in the respective channel, and 0 otherwise.

Signals are sent by an SPE using the API function `spu_sndsig`, which also has variants with fence and barrier effects, see above. The parameters are similar to DMA get and put, but the redundant size parameter is omitted.

The PPU can send signals to an SPE either by using the API function `spe_signal_write` or by accessing the memory-mapped I/O registers in that SPE's MFC directly.

8.3 Cell Programming with the SDK

In this section we present selected key topics in programming Cell using the IBM SDK [28]. For a more comprehensive presentation, we refer for example, to the Cell Programming Handbook [27] and to Scarpino [50].

8.3.1 PPE/SPE Thread Coordination

We demonstrate the basic coordination of SPEs by the PPU with the help of a Hello World program (adapted from [24]). The PPU code is shown in Figure 8.3, the corresponding SPU code in Figure 8.4. Figure 8.5 shows a Makefile for building the program, calling the different gcc compilers for PPU and SPU code.

The SPU executable (here, `spuhello`) can, if statically known, be embedded at compile time as a special data segment into the binary code for the PPU and is thus already in memory when the PPU starts executing, so it only needs to be copied to each SPE from there. For this, the tool `ppu-embedspu` will be used to convert the SPU executable into a PPU object file. This object file, as well as the extended PPU executable resulting from linking it with the actual PPU code, is represented in an extended ELF format that is called *CESOF* (CBE Embedded SPE Object Format).

Here, we apply a second variant, where the SPU executable is kept as a separate ELF file in the file system that is, at run time, looked up and embedded (by memory-mapping) with the running PPU application. In either case, the SPE executable can be referenced by a *handle*, a pointer to an SPE code wrapper structure of type `spe_program_handle_t`. Here, this handle is called `spuprog` (0) and is set by the PPU in the beginning of its `main` function (1).

The PPU then checks the number of available SPEs (2) and determines the number `nspus` of SPE threads to start, one per available SPE. It then creates one context object for each SPE thread (3), calls the system loader to copy the SPU executable into each SPE's local store, and adds a reference to the loaded code to each context object (4).

Now we are, in principle, ready to start SPE computations by calling `spe_context_run` (6). However, that API function is blocking, that is, it returns control to the PPU only after the started SPE computation has terminated. In order to run several SPE computations in parallel (and possibly also let the PPU do something useful at the same time), we have to multithread the PPU program, that is, we create and assign one (POSIX) thread (5) to each SPE context in order to start the SPE computation (7) and wait for it to terminate (9).

The POSIX threads (pthreads) API allows to dynamically create new threads only at function calls; we therefore have to factor out that part into a separate function `ppu_pthread_function`. Once called (5), this function instance will run in its own thread, and a single parameter (`arg`) can be passed to it. Here, we use `arg` to carry a reference to the respective context object for the SPE computation to be started (5,6).

The SPE computation (see Figure 8.4) here simply consists in printing a message (using the PPU's operating system) (8).

```
#include <stdlib.h>
#include <stdio.h>
#include <libspe2.h>
#include <pthread.h>

spe_program_handle_t *spuprog;                              // (0)

#define NUM_SPU_THREADS 6

void *ppu_pthread_function ( void *arg )
{
  spe_context_ptr_t ctx;
  unsigned int entry = SPE_DEFAULT_ENTRY;
  ctx = *((spe_context_ptr_t *) arg );                      // (6)
  spe_context_run ( ctx, &entry, 0, NULL, NULL, NULL );     // (7)
  pthread_exit ( NULL );                                    // (9)
}

int main()
{
  int i, nspus;
  spe_context_ptr_t ctxs[NUM_SPU_THREADS];
  pthread_t pputhreads[NUM_SPU_THREADS];

  spuprog = spe_image_open("spuhello");                     // (1)

  nspus = spe_cpu_info_get ( SPE_COUNT_USABLE_SPES, -1 );   // (2)
  if (nspus > NUM_SPU_THREADS) nspus = NUM_SPU_THREADS;

  for(i=0; i<nspus; i++)
  {
    ctxs[i] = spe_context_create (0, NULL);                 // (3)
    spe_program_load (ctxs[i], spuprog );                   // (4)
    // Create a pthread on PPU for each SPE context:
    pthread_create ( &pputhreads[i], NULL,                  // (5)
                     &ppu_pthread_function, &ctxs[i]);
  }

  // Wait for SPU threads to complete execution:
  for (i=0; i<nspus; i++) {
    pthread_join ( pputhreads[i], NULL );                   // (10)
    spe_context_destroy ( ctxs[i] );                        // (11)
  }
  return 0;
}
```

FIGURE 8.3:   PPU code for the "Hello World" program for Cell. For brevity, all error-handling code was omitted.

```
#include <stdio.h>

int main( unsigned long long speid,
          unsigned long long argp,
          unsigned long long envp )
{
  printf("Hello from SPU 0x%llx\n", speid);              // (8)
  return 0;
}
```

FIGURE 8.4: SPU code for the "Hello World" program for Cell.

```
hello: spuhello hello.c
         gcc -Wall -lspe2 -O3 -o hello hello.c

spuhello: spuhello.c
           spu-gcc -Wall -O3 -o spuhello spuhello.c
```

FIGURE 8.5: Makefile rules for building the "Hello World" program for Cell.

After getting back control from its SPE (8), the controlling PPU thread terminates itself (9). The main PPU thread waits for each PPU thread to terminate (10), after which it can safely deallocate the context objects (11).

8.3.2 DMA Communication

The call to `spe_context_run` (7) can pass only up to three parameters directly to the SPU `main` function (see Figure 8.4). By convention, the first parameter (`speid`) passes the SPE thread ID. The second one (`argp`) often carries over a main memory address of a block of data containing additional parameters (e.g., the main memory addresses of operand blocks or the SPE's rank as viewed by the PPU).

Such a *control block* is usually a data structure such as

```
typedef struct _control_block {
  unsigned int blocksize;    // some size information
  unsigned char pad1[4];     // pad 4 bytes to align at 8 bytes
  unsigned long long addrA;  // address to be accessed by DMA
  unsigned long long addrB;  // address to be accessed by DMA
  unsigned char pad2[8];     // pad 8 bytes to make control block
} control_block;             //      size a multiple of 16 bytes
```

where the main memory addresses of (possibly, dynamically allocated) arrays A, B are then written by the PPU into the control block as follows:

```
control_block cb __attribute__((aligned(128)));  // in main memory
// ...
int *A = (int *) _malloc_align ( ARRAY_SIZE, 7 ); // 128B aligned
```

```
int *B = (int *) _malloc_align ( ARRAY_SIZE, 7 ); // 128B aligned
// ...
cb.addrA = (unsigned long long) ((intptr_t) A );
cb.addrB = (unsigned long long) ((intptr_t) B );
```

Now, the PPU passes cb to the SPE main function as argp parameter:

```
spe_context_run( ctx, &entry, 0, &cb, NULL, NULL );
```

The started SPE program has allocated space for a local copy of the control block in its local store:

```
volatile control_block mycb __attribute__((aligned(128)));
```

The SPE program can then fetch the control block from main memory by a DMA transfer, using the effective address of cb obtained via argp:

```
int tag0 = mfc_tag_reserve();   // get a free tag
if (tag0 == MFC_TAG_INVALID) { ... }
mfc_get( &mycb, argp, sizeof(mycb), tag0, 0, 0 );
mfc_write_tag_mask( 1<<tag0 );
mfc_read_tag_status_all();
printf("Got mycb from main memory\n");
```

Note that the mfc_get call only issues the DMA get command to the SPE's MFC; control returns to the SPE immediately. To make sure that the communication has properly terminated, we have to wait for it, using the tag as a reference: mfc_read_tag_status_all waits for all DMA requests encoded by the MFC tag mask, a bitvector, which is set up by the call to mfc_write_tag_mask and here includes just one DMA tag to monitor, namely, tag0.

Now, the SPE has access to addresses mycb.addrA etc., and can use these in subsequent DMA communication, such as

```
int a[N_BUFFER_ENTRIES] __attribute__((aligned(128)));
int b[N_BUFFER_ENTRIES] __attribute__((aligned(128)));
int tag1, tag2;
// ...
tag1 = mfc_tag_reserve();    // get another free tag
mfc_get ( a, mycb.addrA, sizeof(a), tag1, 0, 0);
mfc_write_tag_mask( 1<<tag1 );
mfc_read_tag_status_all();
compute ( a, b );  // use data in a to compute results in b
tag2 = mfc_tag_reserve();
mfc_put ( b, mycb.addrB, sizeof(b), tag2, 0, 0);
mfc_write_tag_mask( 1<<tag2 );
mfc_read_tag_status_all();
```

to read in the first N_BUFFER_ENTRIES elements from array A into a buffer a in the SPE's local store, use these to compute results in array b, and write back these to main memory by a mfc_put call. If there are more elements to process in A, pointers mycb.addrA and mycb.addrB could now be incremented by sizeof(a), and the get, compute, put sequence could then be repeated for the next bunch of data from A, and so on, until all ARRAY_SIZE entries of A have been processed.

The disadvantage of this simple schema is that much time is wasted in waiting for DMA communication to finish. In the next section, we will show how multi-buffering can be used to overlap DMA communication with SPE computation.

8.3.3 DMA Communication and Multi-Buffering

Multi-buffering of operand and result array elements that are processed in sequence by the SPU (i.e., stream computing) is a technique that allows to overlap DMA communication in time with SPE computation.

A frequent case is *double buffering*, which is shown for the case of an operand array A in Figure 8.6. Two buffers of fixed size are alternatingly used to fetch operands from local store for SPU computations (a1 in the figure) and in parallel refill the other buffer (here, a2) with the next block of operand values. After a1 has been used completely and a2 is refilled, the pointers to a1 and a2 are swapped for the next blocks of A to be computed and fetched, respectively.

Such a pair of buffers is required for each operand array and for each result array of a streaming computation (such as elementwise array processing in a loop). Hence, the number of buffers required is given by the computation structure. Due to the limited size of the local store, this puts an upper limit on size. For complex computations requiring several dozens of buffers simultaneously, this may force us to use relatively small buffer sizes, which means less efficient DMA communication and more control overhead in the SPU code, for example, for switching buffers.

Double buffering is most appropriate when computation and communication require approximately the same amount of time. In certain cases, even triple buffering can be appropriate. Chen et al. [10] describe an algorithm to determine the optimal multi-buffering scheme for given local store space constraints.

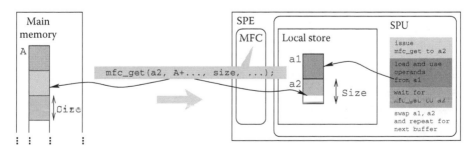

FIGURE 8.6: Double buffering of an operand array with two buffers a1 and a2, containing size bytes each, in the local store.

Data Type	Size Type	#Bytes
bool char unsigned char	Byte	1
short unsigned short	Halfword	2
int unsigned int float	Word	4
double long long(int) unsigned long long(int)	Doubleword	8
qword	Quadword	16

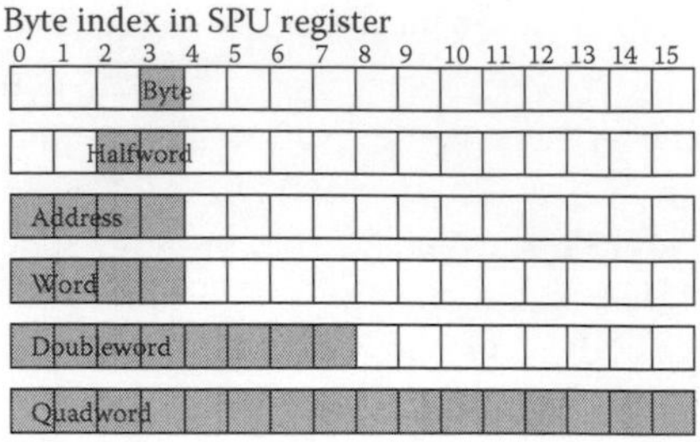

Byte index in SPU register

0　1　2　3　4　5　6　7　8　9　10　11　12　13　14　15

Byte
Halfword
Address
Word
Doubleword
Quadword

FIGURE 8.7: Scalar data types for the SPU with their sizes and mapping to preferred slots in SPU registers.

8.3.4 Using SIMD Instructions on SPE

Basically all SPU instructions work on 128-bit data words. The SPU supports also scalar data types having fewer bytes (see Figure 8.7) but rather as a degenerated case of vector data types, which we will present further below. Scalar data types are stored consecutively in the SPE local store. When a scalar is loaded into a (128-bit) register, it is placed in its so-called *preferred slot*, as shown in Figure 8.7. If necessary, the compiler has to generate special code that extracts the desired bytes from the 128-bit load result and/or shifts these into the preferred slot, and doing vice versa when storing scalar data.

If a scalar is located at a local store address that is not aligned with the scalar's preferred slot in registers, additional shifting is necessary. For instance, the scalar SPE code

```
#define N (1<<14)
float a[N] __attribute__((aligned(16)));
float b[N] __attribute__((aligned(16)));
...
for (i=0; i<N; i+=4)
      a[i] += b[i];
```

for adding $2^{14}/4 = 4096$ scalar floats stored at addresses that are 16-byte aligned, takes 2460 ticks of the SPU high-precision step counter on my PlayStation 3,* where 79.8 ticks correspond to 1 μs, thus 1 tick is 12.53 ns or 40 clock cycles. In contrast, accessing floats not aligned at a 16-byte boundary costs extra time for shifting: The loop

---

* These measurements on PS3 were done using SDK v3.0 with GCC version 4.1.1.

Vector Data Type	Number and Type of Elements
`vector unsigned char`	16 unsigned chars
`vector signed char`	16 signed chars
`vector unsigned short`	8 unsigned shorts
`vector signed short`	8 signed shorts
`vector pixel`	8 unsigned halfwords
`vector unsigned int`	4 unsigned ints
`vector signed int`	4 signed ints
`vector float`	4 floats
`vector unsigned long long`	2 unsigned long longs
`vector signed long long`	2 signed long longs
`vector double`	2 doubles

FIGURE 8.8: Vector data types for the SPU. The element data types are SPU scalar data types (see Figure 8.7). All vector data types are 16 bytes wide.

```
for (i=1; i<N; i+=4)
    a[i] += b[i];
```

takes 2664 ticks. (A closer look at the assembler code reveals that the compiler generated three more instructions per loop iteration.) Worse yet, a misalignment of the two operands, here a[i] at offset 1 and and b[i+1] at offset 2,

```
for (i=1; i<N; i+=4)
    a[i] += b[i+1];
```

causes an additional overhead of two more instructions per iteration, and takes 2766 ticks.

Vector data types refer to a set of 16 consecutive bytes in the SPE local store that are interpreted as a sequence of scalar data types packed into one 128-bit quadword, and will be placed together in a 128-bit register for computations. Figure 8.8 shows the SPU vector data types. Figure 8.9(a) gives an example of a SIMD operation on vector float operands, doing four floating-point additions at a time.

Continuing on our example, vectorization gives an enormous speedup: While adding $2^{14} = 16,384$ consecutive floats by scalar computations

```
for (i=0; i<N; i++)
    a[i] += b[i];
```

takes 9831 ticks, the vectorized version

```
vector float *va = (vector float *)a;
vector float *vb = (vector float *)b;
for (i=0; i<N/4; i++)
    va[i] = spu_add( va[i], vb[i] );
```

takes only 1640 ticks, which is about 6 times faster. This also demonstrates that speedup by vectorization is not limited by the packing factor (here, 4 floats per

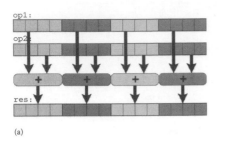
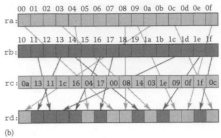

FIGURE 8.9: (a) SPU vector addition (`spu_add`) of two `vector float` values `op1` and `op2`. (b) The shuffle operation `rd = spu_shuffle(ra, rb,rc)`, where the selection vector `rc` indicates the source vector (0 for `ra`, 1 for `rb`) and index position of every byte of the result vector `rd`.

`vector float`), but additional advantage can be drawn from the removal of the masked insertion operations (explained further below) used to store scalars, as now whole 128-bit words can be written in one instruction. In particular, even if we only were interested in every fourth result, it is still 50% faster to compute them all.

The above results were obtained with the `spu-gcc` compiler using the `-O3` optimization level. Additional performance improvements can be obtained, for example, by selecting specific optimization flags in the `spu-gcc` compiler, such as `-funroll-loops` to unroll loops with a number of iterations that is known at compile time or on entry to the loop. Unrolling a loop by a certain number of iterations increases code size but is often profitable, as it reduces the loop trip count and thereby loop control overhead, and it yields more opportunities for subsequent low-level optimizations such as common subexpression elimination and instruction scheduling. In the above example, the scalar loop runs in 8807 ticks and the vectorized version in 1372 ticks after unrolling by a factor of 8. Here, spu-gcc does apparently not exploit the full optimization potential yet, as much better performance can be obtained after manual unrolling by a factor of 8:

```
vector float *va = (vector float *)a;
vector float *vb = (vector float *)b;
for (i=0; i<N/4; i+=8) {
        va[i]   = spu_add( va[i],   vb[i]   );
        va[i+1] = spu_add( va[i+1], vb[i+1] );
        va[i+2] = spu_add( va[i+2], vb[i+2] );
        va[i+3] = spu_add( va[i+3], vb[i+3] );
        va[i+4] = spu_add( va[i+4], vb[i+4] );
        va[i+5] = spu_add( va[i+5], vb[i+5] );
        va[i+6] = spu_add( va[i+6], vb[i+6] );
        va[i+7] = spu_add( va[i+7], vb[i+7] );
}
```

processes 32 floating-point additions per loop iteration and does the job in just 450 ticks. Moreover, applying the `-funroll-loops` compiler transformation to this already manually unrolled code brings the time further down to 386 ticks, which corresponds to slightly more than one float addition result per clock cycle.

Even with a perfectly (modulo-) scheduled loop hiding all instruction latencies and with neglecting all loop control overhead, the SPU could do at most one local store access per clock cycle. Here, two loads and one store instruction are required per quadword, which have to share the SPU's load/store unit (even though they may overlap with vector additions after modulo scheduling, as the floating-point unit and the load/store unit can operate in parallel). Hence, the throughput is here limited by at most four additions per three clock cycles, i.e., $(3 \cdot 16384/4)/40 = 307$ ticks is a lower bound for the execution time. Coming closer to this limit requires more aggressive code optimizations or even assembler-level programming.

There exist many other vector operations for the SPU. A very useful one is the *shuffle operation*, see Figure 8.9b. The operation

```
rd = spu_shuffle( ra, rb, rc );
```

assigns the bytes of the result vector rd from arbitrary bytes of the source vectors ra and rb, guided by the entries in the selection vector rc. Bit 5 in byte i of rc specifies the source vector of byte i of rd, while the four least significant bits in byte i of rc specify the byte position in that source vector that byte i of rd is copied from. Entries > 127 in the selection vector set byte i of rd to special constants; more details can be found in the Cell programming handbook [27]. The selection vector can be determined by other SPU operations such as spu_mask.

A restricted form of the shuffle operation is the *select operation*

```
rd = spu_sel( ra, rb, rm );
```

where the byte positions remain the same (i.e., byte i of rd comes from byte i of either ra or rb) and thus a single bit (bit i) in the mask vector rm is sufficient to select the source vector for each byte position in rd. Again, such bit masks can be created by special vector operations, such as spu_maskb, or by elementwise vector compare operations such as spu_cmpeq (elementwise test for equality) or spu_cmpgt (elementwise test for greater-than). For instance, the following code snippet

```
// a and b are arrays of floats
vector float ra = * (vector float *) a,  // load first 4 elements of a
             rb = * (vector float *) b,  // load first 4 elements of b
             rd;
vector unsigned int rm;   // mask vector
rm = spu_cmpgt ( ra, rb );
rd = spu_sel ( ra, rb, rm );
```

computes the elementwise minimum vector from ra and rb.

The shuffle and select operations are especially useful when extracting or inserting special bytes corresponding to a scalar value in a 128-bit quadword. In this way, a scalar in a register or in local store memory can be modified without modifying the surrounding bytes. Specialized operations `spu_extract` (as used above) and `spu_insert` exist for these frequent cases. In the example above, `spu_extract (rd,2)` will return the third (`float`) element of vector `rd`, that is, the one located at byte positions 08–11.

8.3.5   Summary: Cell Programming with the SDK

The following incremental development process can be used to produce a Cell program under the SDK:

1. Start from an ordinary C program for the PPE only. This program can serve as a baseline version, for instance, for testing purposes and for determining speedups.

2. Identify, maybe by using a profiler, the computationally most demanding parts that could be well suited to offload to one or several SPEs for computation. Very irregular, control-intensive, or I/O-intensive code should remain on the PPE. Move the offloadable code (possibly in several steps) to a single SPE first. Pass the necessary parameters to the SPE in a control block to be fetched by DMA. Add DMA commands to load required operands from main memory to local store before SPU operations, and write results back. Align and pad data structures accessed by DMA operations to meet alignment and size constraints.

3. Use, where possible, SIMD operations to speed up the SPE computation.

4. Add multi-buffering to overlap DMA transfers with SPU computation.

5. Multithread the PPE program and start SPE computations on one or several SPEs. Note that, depending on the problem structure and granularity, using multiple SPEs does not necessarily result in shorter execution time.

The last three steps may, in principle, be done in any order.

With this procedure, the major restructuring steps can be done one by another, and, at least to some degree, applied gradually, to one offload candidate at a time (e.g., more and more functionality can be moved from PPE to SPE, starting with the computationally heaviest kernels), such that there exist many points during development where a complete version of the program is available and can be tested against the baseline or previous version. This incremental development process is recommended because SDK programming is a quite error-prone and time-consuming process.

8.4   Cell SDK Compilers, Libraries, and Tools

In this section, we give a short and nonexhaustive overview of existing tools, libraries, and frameworks for Cell programming provided in the IBM SDK [28]. High-level programming environments for Cell will be discussed in Section 8.5.

8.4.1 Compilers

Gschwind et al. [18] describe the open-source development environment for Cell based on the GNU toolchain, which forms the basis of the Cell SDK. This includes a port of the GCC compiler, assembler, and linker, of the GDB debugger, but also many other useful tools such as the embedding tool for SPE executables in PPE executables.

XL C/C++ is a proprietary compiler by IBM that was ported to Cell by Eichenberger et al. [14]. XL C provides advanced features such as auto-vectorization, auto-parallelization, and OpenMP support (see also Section 8.5.1). The XL C compiler can be used together with the other, open-source based development tools for Cell, such as the GDB debugger.

A recent GCC (v4.3) based compiler for Cell that includes improved support for a shared address space, auto-vectorization, and automatic code partition management is described by Rosen et al. [48].

OpenCL [39] is a recently developed open standard API for heterogeneous and hybrid multicore systems. Since 2009, IBM offers an implementation of the XL compiler for OpenCL [29] for the most recent generation (QS22) of its Cell blade servers.

8.4.2 Full-System Simulator

IBM full-system simulator for Cell allows to simulate Cell-based systems on Linux platforms. The simulator can be configured in architectural parameters such as the number of SPEs or the simulated system memory size. The simulator also contains a mode for cycle-accurate simulations, which however makes simulation slow.

8.4.3 Performance Analysis and Visualization

The cpc (Cell Performance Counter) tool allows to collect global performance counters on a running Cell program during a specified time period, which provides statistics about clock cycles, branches, cache misses, and other hardware events on the various hardware units of Cell.

OProfile is a system-level profiler for Cell.

For the SPEs, there exists the *SPU timing tool*, a static pipeline visualizer that annotates the SPU program's assembler code with a graphics that shows dispatch rate, pipeline stalls, and other information about the performance of instruction-level parallel execution in the SPU.

The *Performance Debugging Tool* (PDT) allows for instrumentation of Cell code in order to trace events of interest of a running Cell program.

VPA is a graphical user interface for performance visualization based on Eclipse. It integrates tools for profile analysis, trace analysis, code analysis, pipeline analysis, performance counter analysis, and control flow analysis.

The performance analysis tools are optional extensions to the SDK.

8.4.4   Cell IDE

The Cell IDE (Integrated Development Environment), implemented on top of the Eclipse CDT, is an optional extension of the SDK.

8.4.5   Libraries, Components, and Frameworks

The IBM Cell SDK provides several libraries with optimized implementations of mathematical functions on vectors and arrays. These include the SIMD Math library, the MASSV library, libraries with common operations on geometric data, for example, point transformations using 4×4 matrices, a library for solving linear equation systems represented by large matrices, and an implementation of the Basic Linear Algebra Subroutines (BLAS) [22]. Also, a library for Fast Fourier Transform (FFT) computations is provided.

SIMD Math library: The SIMD Math library [25] provides an extension to the SPU SIMD intrinsics, that is, functions that operate on 128-bit SPU `vectors`. The SIMD Math library also provides inlineable versions of its functions that allow to skip the call overhead for invoking these, and fast versions that are designed for execution speed rather than numerical accuracy.

MASSV library: The Mathematics Acceleration Subsystem for Vectors (MASSV) library provides data parallel mathematical functions that operate on `float` arrays in local store with an arbitrary number of elements (i.e., not on SPU `vectors`). These functions are optimized for the SPU, e.g., by using SPU vector instructions. Using MASSV functions leads generally to faster code compared to equivalent loop constructs over corresponding SPU SIMD intrinsics or SIMD Math function calls.

ALF: The Accelerated Library Framework (ALF) [26] is a platform-independent API for developing task parallel and data parallel applications on accelerator-based multicore systems. The SDK contains an implementation of ALF for Cell.

The programmer first identifies *code units* that could run on an accelerator (here, SPE), such as libraries or user code modules of computationally intensive functions; these functions are referred to as *computational kernels*. The programmer partitions the input and output data structures (usually, large arrays) to be processed by a kernel invocation into *work blocks*. The invocation of a computational kernel for one work block thus forms a *task*, that is, a unit for dynamic scheduling and dispatching to an SPE by ALF.

Code units can be mapped to specific accelerators (SPEs) for execution. ALF supports both *SPMD* (single program, multiple data streams) and *MPMD* (multiple programs, multiple data) execution. In SPMD execution, the same code unit is made available on each SPE and hence all SPEs could participate in the data parallel computation of the same computational kernel's invocations, this is appropriate for computationally intensive data parallel computations on partitionable large data structures such as arrays; each SPE works on a different partition of the data. MPMD execution is appropriate for task-parallel problems, where invocations of different computational kernels can be executed at the same time on different SPEs.

The code for the master processor (PPE) contains API calls that sets up *task descriptors* for kernel invocations to be launched to the accelerators (SPEs). Each task descriptor is filled with information about the code units and kernels to be invoked, lists of input and output parameters of a work block to be transferred to the SPE, SPE buffer sizes for input and output parameters, SPE stack size, and some other metadata. Next, the tasks are registered with the ALF runtime system and enqueued in a global ALF task queue, from where they are scheduled to local work queues of the SPEs. The default task scheduling policy of ALF can be reconfigured.

The programmer can also register dependences between tasks of different code units, which constrain the dynamic task scheduler in the ALF runtime system.

The code units themselves (i.e., SPE code) need also some calls to the ALF API to control the life cycle of the code unit and its kernel invocations with their parameters and runtime data. There are up to five different code parts (*stages*) in a code unit to be marked up by the programmer: for setup of the computational kernel code, setup of DMA communication lists and buffer management for input parameters, of an invocation (task), control of the actual computational kernel, setup of DMA communication lists and buffer management for output parameters, and postprocessing in the code unit. Each stage is marked by an ALF macro.

ALF then takes care of the details of task management, splitting work blocks into packets that fit the specified buffer sizes in local store, and multi-buffered DMA data transfers of operand packets to and from the executing SPEs automatically.

In general, ALF code is less complex and error-prone than plain-SDK code, especially because DMA communication and multi-buffering need no longer be coded by hand. However, the SPE code for the computational kernel still needs to be vectorized and optimized separately. On-chip (SPE-SPE) pipelining for tasks of different code units is not supported in ALF.

8.5 Higher-Level Programming Environments for Cell

SDK-level programming gives the programmer close control of the Cell hardware resources. However, this comes at the price of a low level of abstraction, which in turn means lengthy and error-prone code and longer development times.

In the following, we present some higher-level programming environments for Cell that could be used as alternatives to increase programmer productivity—usually, at the cost of reduced code efficiency compared to expert-written SDK code. They abstract from one or several aspects of SDK programming. For instance, almost all of them provide a global address space abstraction, and they often restrict the many ways of how Cell programs could be organized to compositions of only a few simple constructs, such as data parallel loops or reductions, which should be used instead of their lower-level SDK counterparts. More advanced parallel coordination patterns such as on-chip pipelining between SPEs may not always be expressible efficiently with these programming frameworks. However, for more regular scenarios they can be useful.

The list of Cell-specific high-level programming environments mentioned in this section is in no particular order and by no means exhaustive. For a more comprehensive survey, classification, and comparison of programming environments for Cell, we refer to Varbanescu et al. [56].

8.5.1   OpenMP

The Octopiler compiler by Eichenberger et al. [14], based on the IBM XL C/C++ compiler for the PowerPC architecture, provides an OpenMP implementation for Cell that couples the SPEs and the PPE together to a virtually shared memory multiprocessor with a uniform OpenMP programming interface. It automatically creates PPU and SPE code files from a single OpenMP source program, where the SPE code only contains those code parts that should be executed in OpenMP `parallel` regions on the SPEs. By embedding the SPE code into a PPE data segment linked with the PPE code, a monolithic executable file is produced.

A major part of the local store of each SPE is set up to act as an automatically managed *software cache* to provide an abstraction of a global shared memory, which has its home residence in main memory. The software cache functionality is provided by a library for shared memory accesses that can be linked with the SPE code. Accesses that cannot be handled by the cache (i.e., cache misses) lead to DMA operations that access the corresponding locations in main memory. There exist two variants of the software cache: one variant that only supports single-threaded code, where any cache line is written back completely to memory when evicted at a cache miss, and a second variant that supports multithreaded shared memory code (as in OpenMP applications). The latter variant keeps track of modified cache lines such that, either on eviction or at write-backs enforced by the consistency mechanism (e.g., at OpenMP `flush` operations), only the modified bytes are written back by DMA operations.

The software cache implementation uses SIMD instructions for efficient lookup of shared memory addresses in the local store area dedicated to holding software cache lines. By default, the Octopiler instruments all accesses to shared memory (in code intended for SPE execution) to go via the software cache's lookup function. The resulting code is usually not efficient yet. In order to save some lookup overhead for accesses that are statically expected to yield a cache miss, the compiler can choose to bypass the cache lookup and access the value directly from its home location in main memory. Prefetching and coalescing of memory accesses are other optimizations to improve the software cache performance. Such optimizations have been shown to boost the speedup achievable with the software cache considerably [14].

The Octopiler performs automatic vectorization of loops to exploit SIMD instructions, and performs optimizations of the alignment of scalar values and of entire data streams to improve performance.

Large SPE executables constitute a problem for Cell because of the very limited size of the local store (or, respectively, the part of the local store that is not reserved for, e.g., the software cache and other data). The Octopiler supports automatic splitting of larger SPE programs into partitions and provides a runtime partition manager that (re)loads partitions into the local store when they are needed. Only a few partitions can be held simultaneously in the local store at a time. Cross-partition calls and branches are therefore replaced by stubs that invoke the runtime partition manager to check if the target partition is currently resident in the local store, and if not, reload them before the SPE program can continue execution. If necessary, another loaded partition must be evicted from the local store instead.

With selected benchmarks suitable to the Cell platform, average speedups of 1.3 for SPE-specific optimizations (such as branch optimizations and instruction scheduling), 9.9 for auto-vectorization, and 6.8 for exploiting thread-level parallelism in OpenMP programs were obtained with Octopiler [14].

DBDB [42] and Cellgen [51] are implementations of OpenMP, respectively, an OpenMP subset that do not rely on a software cache for the shared memory abstraction but use instead a combination of compiler analysis and runtime management to manage data locality.

Extensions of the OpenMP (3.0) task model that should allow to better address heterogeneous multicore architectures such as Cell have been proposed [5]. For instance, device-specific (e.g., SPE) code for an OpenMP (3.0) task can be integrated as an alternative implementation variant to the portable default (i.e., CPU-based) implementation.

8.5.2   CellSs

Cell superscalar (CellSs) [6] provides a convenient way of exploiting task-level parallelism in annotated C programs automatically at runtime. Starting from an ordinary C program that could run on the PPU only, the programmer marks certain computationally intensive functions as candidates for execution on SPEs and declares their interface with parameter types, sizes and directions. CellSs builds, at runtime, a task graph of tasks that could run on SPEs in parallel with the PPU main thread, keeps track of data dependences, and schedules data-ready tasks at run-time to available SPEs.

CellSs is implemented by a source-to-source compiler and a runtime system. The source-to-source compiler generates SPE code for these functions, which is further processed by the SPE compiler tool chain, as well as special stub code for all (PPU) calls to such functions. When, during program execution, the PPU control hits such a stub, the runtime system creates a new task node in the task graph to represent the call with its parameters, detects data dependences of the new task's input operands from previously created (and yet unfinished) tasks, and represents these as edges in its task graph structure. The stub call is not blocking, i.e., PPU control continues in parallel to this. The runtime system schedules the data-ready tasks dynamically to SPUs. Dynamic task graph clustering for reducing synchronization overhead by increased granularity, parameter renaming to remove write-after-read and write-after-write dependences, locality-preserving mapping heuristics, and task stealing for better load balancing have been added as optimizations to the basic mechanism. Overall, this method is similar to how a superscalar processor performs dynamic dependence analysis, resource allocation, mapping, and scheduling on a sequential instruction stream.

For exploiting the other types of parallelism in Cell (such as automatic vectorization for SIMD parallelism and automatic multi-buffering for DMA parallelism), CellSs relies on an underlying Cell compiler tool chain such as Octopiler (see Section 8.5.1) to provide these. Task operands in CellSs are restricted to scalars and arrays.

8.5.3 Sequoia

Sequoia [16] is a programming language for Cell and other parallel platforms that takes a slightly more declarative approach than most other programming environments discussed here. For instance, Sequoia explicitly exposes the main structure and cost of communication between the various memory modules in a system in the form of a tree representation of the memory hierarchy.

Sequoia provides tasks, which have a private address space; the only way of communicating data between tasks (and thus possibly between cores) is restricted to invocations of subtasks by parent tasks, i.e., operand transfer at remote procedure calls. Sequoia offers three skeleton-like primitives to create subtasks, namely, a parallel multidimensional forall loop, a sequential multidimensional loop, and *mapreduce*, a combination of a parallel loop and subsequent tree-like parallel reduction over at least one operand.

Sequoia defines special operators to split and concatenate arrays, which support, for instance, parallel divide-and-conquer computations on arrays such as recursive matrix-matrix multiplication or FFT.

The programmer may define multiple algorithmic variants for solving the same task. Moreover, these can be specialized further to expect their operands to be present in certain memory modules. Selecting among these variants (if necessary, including moving data between memory modules) is a way to optimize execution time. Also, certain variables (e.g., those that control how to split a problem into subproblems) can be defined as *tunable parameters* so that the computation can be optimized for a given target machine. The selected values for these parameters are specified externally for each algorithmic variant in a machine-specific mapping specification.

An implementation of Sequoia, done at Stanford university, is available for the Cell SDK 2.1 (2007).

8.5.4 RapidMind

RapidMind Multicore Development Platform by RapidMind Inc. [44] (which was bought by Intel in 2009) is a stream programming language that provides a data parallel programming extension for C++. RapidMind defines special data types for values, arrays, and functions. Data parallelism is specified by applying a componentized scalar function elementwise to its input arrays and values. If the function is so simple that its execution time can be assumed to be independent of the input data, the data

```
int main()
{   ...
    int x = 0;      // x is global to the sieve block
    sieve {
        int y = 1;  // y is local to the sieve block
        x = y + 1;  // writes 2 to x at the end
        y = x;      // sets y to 0 as the new value of x is not visible yet
    }               // here the write to x takes effect
    ...
}
```

FIGURE 8.10: Example of a sieve block in Sieve C++.

parallelism can be scheduled statically across the SPEs. For more complicated functions where execution time may vary, dynamic load balancing can be added. Also, the data parallel function specification enables the automatic selection of SPE SIMD instructions and automatic application of multi-buffering to the operand arrays.

From the same RapidMind source program, code for Cell as well as for GPUs and other data parallel platforms can be generated without modification, thanks to the simplicity and universality of the data parallel programming model.

8.5.5 Sieve C++

Static analysis of C/C++ code for proving independence of certain statements is generally difficult due to the potential aliasing of program variables, for instance via pointers, or via array index expressions that are hard to analyze or depend on runtime data. With insufficient static information about aliasing of accessed variables, the existence of data dependences between statement instances must be conservatively assumed, which can prohibit automatic parallelization, loop vectorization and other useful code optimizations that would modify the original (sequential) execution order.

Sieve C++ [43] is a language extension and modification to C++ that should alleviate the automatic extraction of parallelism. The main new language construct is the sieve construct. Sieve blocks usually mark code parts that are candidates for offloaded execution on (programmable) accelerators such as Cell's SPEs.

In a sieve block, all write accesses to variables declared globally to that block are delayed and take effect at the end of the block's execution. For instance, in the example code in Figure 8.10, the write access to x is delayed to the end of the sieve block.

Consequently, of possibly several assignments to the same block-global variable in a sieve block, only the last one would really take effect. Note that delayed updating changes the standard program behavior (namely, sequential consistency) to a more relaxed (yet deterministic) memory model. The point is that there cannot exist data races due to both reading and writing accesses to block-global variables within a sieve block, whether through aliased variables or not. Sieve code can therefore lead to reduced constraints for an automatically parallelizing and vectorizing compiler.

Special data types for parallel iteration variables and parallel reduction variables allow the programmer to convey additional information about independence of computations to the compiler.

Sieve C++ has been implemented by Codeplay [47] for Cell, for x86-based standard multicore systems and for GPUs. The Sieve C++ compiler analyzes the remaining data dependences (on block-local variables) in each sieve block and splits the sieve body into independent tasks for parallel execution.

8.5.6   Offload C++

Codeplay recently released *Offload C++* [11], an extension of C++ for heterogeneous multicore processors, which was primarily designed and implemented for Cell but also aims for portability across a larger range of architectures.

The central new construct in Offload C++ is the offload construct, which marks a code block (the *offload block*) to be spawned off for execution on an SPE. Code outside offload blocks will be executed by the PPE. The offload construct is available in an asynchronous (__offload{...}) and a synchronous (__blockingoffload{...}) variant. The latter requires the PPU thread to wait for the spawned SPE computation to terminate its offload block, as in the example of Figure 8.11. In contrast, asynchronous offloading spawns the SPE thread and immediately returns control to the PPE together with a handle to the spawned SPE computation, which can later be used for synchronization, similarly to our earlier SDK example in Figure 8.3. This allows the PPE to continue and possibly spawn further SPE computations in subsequent offload blocks.

Offload blocks can be parameterized. This feature allows to perform work division in the sense of a parallel for construct [12] by iteratively spawning multiple SPU computations from the same offload block with asynchronous offloading, each being assigned a sub-range of the data to be processed.

The Offload C++ compiler creates two versions of each function called from within an offload block, one compiled for the PPU and one for the SPU. The PPU version is used at calls in outer (i.e., PPU) code while calls from within offload blocks are directed to the SPU version.* The dynamic extent of an offload block (i.e., including the SPU versions of all functions called from it directly or transitively) will thus be executed by an SPU; it is referred to as an *offload scope*.

Unless explicitly permitted by setting a compiler flag, the offload construct cannot be nested, i.e., it is not admissible in an offload scope. Functions declared with the function-type qualifier __offload can only be called from within offload blocks and may contain SPU-specific code. An __offload function can coexist with a (PPU) function of the same name, such as bar in Figure 8.11, where the SPU version is an explicitly defined specialization that will be automatically chosen when called in offload scopes.

---

* There are a few more issues and exceptions from this rule, e.g., regarding function pointers and virtual functions, but we cannot go into details here and refer to [11] for further description.

```
void foo( void ) {...} // compiler creates one PPU and one SPU version

void bar ( void ) {...} // here used in PPU calls only
__offload void bar(void){...} //SPU-only, may contain SPU-specific code

int main( void )    // program entry executed by PPU
{
    int x = ...;    // PPU variable, allocated in main memory
    int *px = ...;    // PPU pointer variable
    foo(); bar();    // calls to PPU versions of foo and bar
    __blockingoffload {  // spawn this offload block to some SPU
        int y;  // SPU variable in local store
        int *py;  // SPU pointer variable to SPU address
        __outer int *pz; // SPU pointer variable to main memory address
        y = x;  // implicit DMA: assigned value x copied in from main memory
        pz = px; // OK, pointees' memory types match
        foo();    // call to SPU version of foo
        bar();    // call to specialized (SPU) version of bar
    } // PPU waits here for spawned SPU computation to finish
    ...
}
```

FIGURE 8.11: Example of a synchronous offload block in Offload C++.

Variables defined inside an offload scope are by default allocated in the executing SPE's local store, while those defined in the outer scope (PPU code) reside in main memory. Accesses from an offload scope to variables residing in main memory result in DMA transfer of the value to be read or written; the implementation [11] uses a software cache to reduce the number of DMA accesses.

Hence, when declaring pointer variables in an offload scope, the memory type of their pointee must be declared as well, by adding the __outer qualifier for pointers to main memory, (unless the compiler can deduce this information automatically), see the example in Figure 8.11. The pointee's memory type thus becomes part of the pointer type and will be checked statically at assignments etc. We refer to [11] for further details.

By providing preprocessor macros that wrap all new constructs in Offload C++ and that could alternatively expand to nothing, portability to other compilers can be achieved. The option of SPU-specific restructuring of offload code allows to trade higher performance on Cell for reduced portability. The portability problem can be avoided, though, by conditionally overloading a portable routine with an SPU-specific version that is masked out when compiling for a non-Cell target. Templates can be used to combine portable data access with special code for optimized DMA transfer in offload functions [12].

Recent case studies of using Offload C++ for Cell include parallelizing an image filtering application [13] and porting a seismic wave propagation simulation code that was previously parallelized for Intel Threading Building Blocks [12].

8.5.7 NestStep

NestStep [35,36] is a C-based partitioned global address space language for executing *bulk-synchronous parallel (BSP)* programs with a shared memory abstraction on distributed memory systems. Similar to UPC (Universal Parallel C [15]) and its predecessors, NestStep features, for example, shared variables, blockwise and cyclically distributed shared arrays, and dataparallel iterators. In contrast to UPC, NestStep enforces bulk-synchronous parallel execution and has a stricter, deterministic but programmable memory consistency model; in short, all modified copies of shared variables or array elements are combined (e.g., by priority commit or by a global reduction) at the end of a BSP superstep to restore consistency, in a way that can be programmed individually for each variable. This combine mechanism is integrated in the barrier synchronization between subsequent supersteps.

As an example, Figure 8.12 shows a superstep with a dot product computation in NestStep, where each SPE calculates a local dot product on its owned elements of A and B and writes its local result to its local copy of the replicated shared variable s. In the subsequent (implicit) communication phase of the superstep, these written copies are combined by summing them up (<+>), and the sum is committed automatically to each copy of s, in order to restore the invariant that all copies of shared variables have the same value on entry and exit of a superstep.

Both NestStep's step statements and Sieve C++'s sieve blocks delay the effect of writes to block-global shared variables and thereby guarantee absence of block-local data dependences on shared variables. The difference is that parallelism in NestStep is explicit (SPMD) while it is implicit (automatic parallelization) in Sieve C++. Also, write accesses are committed locally in program order in NestStep.

NestStep was originally developed and implemented for cluster systems on top of MPI [35]. More recently, the NestStep runtime system was ported to Cell to coordinate BSP computations on a set of SPEs, where the data for each SPE, including its owned partitions and mirrored elements of distributed shared arrays, are kept in

```
sh float s;        // replicated shared variable - one local copy per SPE
sh float A[N]</>;  // block-wise distributed shared array
sh float B[N]</>;  // block-wise distributed shared array
int i;             // private variable
...
step {             // BSP superstep, executed by a group of SPEs
   s = seq_dot_product( owned(A), owned(B), sizeof_owned(A) );
   ...
}
combine ( s<+> );
```

FIGURE 8.12:   NestStep example: Bulk-synchronous parallel dot product computation. The routine seq_dot_product can be implemented either as platform-specific SPE code with intrinsics, or as library function call or as a BlockLib (see Section 8.5.8) skeleton instance.

a privatized area of main memory, and the PPU is used for SPE coordination and synchronization [32].

NestStep for Cell only addresses thread-level parallelism across SPEs and provides a global shared address space abstraction, while support for SIMDization and multi-buffered DMA must be provided separately, either as hand-written SPE code, or in a platform-independent way by using BlockLib skeletons (Section 8.5.8).

8.5.8 BlockLib

BlockLib [2] is a skeleton programming library that aims to make Cell programming simpler by encapsulating memory management, doubly-buffered DMA communication, SIMD optimization and parallelization in generic functions, so-called *skeleton functions*, that are parameterized in problem-specific sequential code. Block-Lib provides skeleton functions for basic computation patterns. Two of these are map and reduce, which are well known. BlockLib also implements two variants of map and reduce: a combined mapreduce, where a reduce operation is applied immediately to the result of a map, and a map-with-overlap for calculations on array elements that also access nearby elements. The library consists of compiled code and macros and requires no extra tools besides the C preprocessor and compiler.

BlockLib is implemented on top of the NestStep run-time system for Cell [32], from which it inherits the data structures for distributed shared arrays and some synchronization infrastructure. However, it could also be used stand-alone with minor modifications. By default, a call to a BlockLib skeleton function constitutes a Nest-Step superstep on its own, but the map skeleton can also be run as part of a larger superstep.

The parameterization in problem-specific user code can be done in different ways that, on Cell, differ very much in performance and ease of use. For instance, generic functions with fine-grained parameter functions (i.e., one call per element computation) are convenient but incur too much overhead on the SPEs and are not amenable to SIMD code generation either. Hence, BlockLib expects user functions with larger granularity here. BlockLib also provides the user with the power of SIMD optimization without the drawbacks of doing it by hand. A simple *function definition language*, implemented as C preprocessor macros, allows to generate SIMD optimized inner loops. This macro language provides primitives for basic and advanced math operations. It is easy to extend by adding definitions to a header file. Many of these macros have a close mapping to one or a few Cell SIMD instructions, and some are mapped to functions in the IBM SIMD Math library [25].

Using BlockLib does not tie the user code to Cell. The same interface could be implemented by another library for any other NestStep platform in an efficient way, with or without SIMD optimization.

BlockLib synchronization is based on Cell *signals*. A signal is implemented with a special register in each of the SPE's MFC. An SPE sends a signal to another SPE by writing to the other SPE's signal register. In BlockLib, the signal register is used in or-mode, which means that everything written to the register is bitwise or-ed together.

This way multiple SPEs can signal the same target SPE without overwriting each other's signals. BlockLib uses one of the two signal registers per SPE. As a signal register is 32 bit wide, each SPE can have its own bit with exclusive write access for up to four different kinds of signals in all other SPEs' signal registers. BlockLib uses internally three kinds of signals (barrier synchronization, message available, and message acknowledge).

A skeleton library for Cell based on C++ templates, called SKELL BE, was recently proposed by Saidani et al. [49].

8.5.9   StarPU for Cell

StarPU is a runtime system for heterogeneous multicore systems such as Cell [4] and GPUs. StarPU provides API functions that enable the programmer to denote offloadable tasks (so-called *codelets*), to expose implementation variants for different core types, and to register codelet operand data stored in special container objects such that StarPU can keep track of operand data status and schedule data-ready tasks at runtime. StarPU keeps history data about task execution times depending on operand sizes, which can be used to predict execution times for future invocations and thus select and dispatch the (predicted) fastest implementation variant at runtime [3].

8.5.10   Other High-Level Programming Environments for Cell

Beyond the ones described earlier in this section, further academic and commercial high-level programming environments for Cell have been developed in the last years. These include:

- SP@CE [55], a streaming programming environment for consumer electronics applications. SP@CE is based on the Series-Parallel Contention (SPC) programming model introduced by van Gemund [53], where synchronization patterns are restricted to series-parallel process graphs.

- MCF [7], the MultiCore Framework by Mercury Computer Systems Inc. The MCF programming model is restricted to data parallel computations on n-dimensional matrices.

- CHARM++ for Cell [41], an implementation of the C++-based, distributed-memory parallel programming model CHARM++ [33] on Cell, extended by an API that allows the PPE program to off-load specific (compute-intensive) tasks to SPEs.

- MPI microtask: With its eight SPEs, each having its own local memory module and communicating by DMA transfers, Cell can be regarded a distributed memory message passing system. IBM's MPI microtask [45] is an implementation of the widely used Message Passing Interface (MPI) that, in principle, allows Cell programmers to write ordinary MPI programs, given that these are broken down into *microtasks*, small-footprint MPI processes that each fit into the local store completely (including code and all data). The decomposition is done by some additional API functions that allow to create groups of micro-tasks dynamically and provide communication in between these.

Moreover, there exist domain-specific APIs and programming environments for Cell, in particular for graphics and game development.

We do not go into further details here and refer the interested reader to the literature instead.

8.6 Algorithms and Components for Cell

There exists already a wealth of algorithms for typical problems, for example, in scientific computing, signal and image processing, that have been especially tailored for implementation on Cell. Due to limited space, we can only give a small selection here.

Scientific Computing: Williams et al. [58,59] investigate implementations of scientific computing kernels such as dense matrix-matrix multiplication, sparse matrix vector multiplication, stencil computations, and 1D and 2D FFT on Cell and demonstrate a large potential for speedup and power-efficiency compared to superscalar processor architectures.

For computational kernels of a bioinformatics application doing multiple sequence alignment, Vandierendonck et al. [54] report a speedup of up to 51.2 on Cell over PPU-only execution.

Sorting: Sorting on Cell is challenging for several reasons: (1) sorting requires much data movement but little computation, (2) the small size of the local store enforces the use of external memory algorithms for sorting on SPEs; and (3) the SPEs do not perform well on irregular, branch-intensive code that is typical for many sorting algorithms such as quicksort, because it is hard to vectorize and leads to many pipeline stalls.

Sorting networks such as mergesort or bitonic sort meet these expectations, because they support external sorting, have a simple, data-independent control structure and can be formulated to efficiently use SIMD instructions. See, for example, JáJá [31] or Akl [1] for a survey of parallel sorting algorithms.

CellSort [17] and AAsort [30] are sorting algorithms for Cell based on bitonic sort and mergesort. Generally, they work in two phases: In phase 1, the SPEs sort blocks of size m that still fit in the local store. In phase 2, these sorted blocks are combined to a fully sorted data set, either by monotonic or bitonic merging. As (serial) merging has a constant in-degree (typically, 2 or 4 streams are merged together at a time), phase 2 is organized as a tree of stream merge operations that produce longer and longer sorted sequences, where each merger (2-way or 4-way) is implemented using SIMD instructions and uses double buffering to overlap loading the next packets of the input streams with merging and storing merged packets back to main memory.

The stream mergings in this second phase still requires $O(n \log(n/m))$ memory accesses to sort n elements from presorted blocks of size $m < n$, and the reported speedups for phase 2 are small. For instance, in AAsort [30], mergesort with 4-to-1-mergers is used in phase 2, where the mergers use bitonic merge locally. The tree of mergers is processed level-wise in rounds. As each SPE reads from main memory and writes to main memory (*dancehall organization*), all n words are transferred from and to main memory in each round. Speedup still is limited, as the main memory interface bandwidth is the performance bottleneck.

Keller and Kessler [34] propose *on-chip pipelining*, a technique to trade a reduced number of memory accesses for an increased volume of SPE–SPE forwarding via the EIB, which has considerably higher bandwidth. Instead of processing the merger tree level-wise to and from main memory, all mergers at all levels run concurrently as dynamically scheduled tasks on the SPEs and forward merged packets immediately to their successors in the merger tree. The merger tree root is the bottleneck in merging and gets an SPE of its own, while the remaining merger tasks are mapped to the remaining SPEs to maximize throughput and minimize the maximum number (thus maximize the size) of buffers used per SPE, which translates into improved DMA efficiency and reduced scheduling overhead. Optimal or optimized mappings can be computed automatically with different algorithms [34,37]. An example mapping for a 32-to-1 merger tree is shown in Figure 8.13.

By on-chip pipelining with suitable mappings, the global merging phase, which is dominating the time of parallel mergesort for reasonably large (8M or larger) input data sets, achieves a speedup over that of CellSort by up to 70% on an IBM QS-20 dual-Cell blade server and by up to 143% on a PS3. Implementation and evaluation details are described by Hultén [20,21].

The on-chip pipelining technique can also be applied to other memory-intensive dataparallel computations [38].

Image and signal processing: The IBM SDK contains an optimized library for fast Fourier transform (FFT) on Cell in an extension package. The application of the Spiral autotuning approach to optimized FFT program generation for Cell is described by Chellappa et al. [8].

H.264 video coding/decoding on Cell has been described, for example, by Wu et al. [60,61]. The use of Cell for real-time video quality assessment was presented by Papp et al. [46].

In a case study, Varbanescu et al. [57] evaluate different approaches for mapping an application for multimedia analysis and retrieval to Cell, and derive general parallelization guidelines, which could extend the process described in Section 8.3.5.

8.7 Summary and Outlook

Four different kinds of parallelism need to be exploited and coordinated properly for efficient Cell programming:

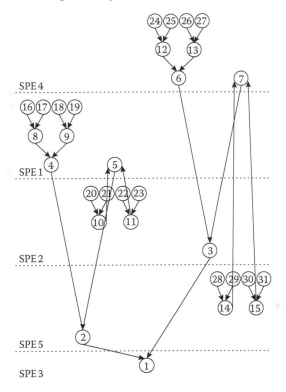

FIGURE 8.13: On-chip pipelined mergesort, here using an optimized mapping of the 31 merger tasks of a pipelined 32-to-1 merger tree to five SPEs of Cell. This mapping was computed by integer linear programming [34] and is optimal w.r.t. computational load balancing as primary optimization goal, the maximum number of buffers needed on any SPE as secondary goal, and the overall inter-SPE communication volume as tertiary goal. All merger tasks mapped to an SPE are scheduled round-robin by a user-level scheduler; tasks that are not data ready are skipped.

- Instruction-level parallelism

- SIMD parallelism

- Memory (DMA) parallelism

- Thread-level parallelism across SPEs and possibly also the PPE

While tools and high-level programming environments can help with some of these, writing high-performance code orchestrating all of these nicely together is still mainly the programmer's task.

Cell is a fascinating processor. In fact, it has created its own class of moderately heterogeneous and general-purpose multicore processors that is clearly set apart from standard multicore processor designs, but also differs from even more domain-specific chip multiprocessor architectures.

With general-purpose GPUs such as NVIDIA's CUDA-based devices described in the previous chapter of this book, Cell shares several architectural features, such as the concept of programmable accelerators that are optimized for a specific type of computation (fixed-width SIMD for Cell SPEs, massively data-parallel for GPUs), and a (mostly) software-managed multilevel memory hierarchy. Both have a more complicated programming model but also a much higher performance per watt potential than comparable standard multicore processors. Both have much higher peak bandwidth for accesses to on-chip than off-chip memory. Both architectures perform better at regular, bulk access to consecutive memory locations compared to random access patterns; while bulk access is explicit (by DMA) in Cell, it is implicit in GPUs (by coalescing in hardware in CUDA GPUs). Also, both platforms recently added hardware support for high-throughput double precision arithmetics, targeting the high-performance computing domain. However, there are also significant differences: While Cells SPEs are powerful (at least in terms of year 2006 technology when Cell was new) general-purpose units with moderate-width (128 bit) SIMD support and a comparably large amount of on-chip local memory, GPUs' streaming processors are much simpler scalar devices that, until recently, did not support function calls in kernel code (unless statically inlined by the compiler); the latter restriction was relaxed in NVIDIA's Fermi GPU generation introduced in 2010. With the PPU, Cell has a general-purpose host processor directly on chip that runs a full-fledged standard operating system, while the entire GPU is just a complement to an external CPU. Both PPU and SPUs can initiate DMA operations to access any off-chip or on-chip memory location (as far as permitted within the current process), while data transfer between GPU memory and main memory can, up to now, only be initiated by the host. In particular, SPEs can forward data to each other, which is not (yet?) possible on GPUs, making advanced techniques such as on-chip pipelining not applicable there. Task parallelism is naturally supported by Cell's MIMD parallelism across its eight SPEs (each of which could even run a different binary code). Until recently, CUDA GPUs could only execute a single, massively dataparallel kernel across the entire GPU at a time; this has now changed with Fermi, which allows for limited task-level parallelism. By leveraging massive hardware parallelism across hundreds of simple streaming processor cores, GPUs still can afford a lower clock frequency. GPU architectures are designed for high throughput by massive hardware multithreading, which automatically hides long memory access latencies, while SPEs require manual implementation of multi-buffering or other latency hiding techniques in software to achieve high throughput. To keep pace with the recent development of GPUs but also of standard multicore architectures, a successor generation in the Cell processor family would have been urgently needed.

In late 2009, IBM decided to cancel further development of the Cell architecture line (earlier there had been some plans for a 2-PPU, 32-SPU version), but will continue to manufacture Cell chips for the PS3 for the foreseeable future. Hence, PS3 and existing installations such as IBM blade servers and the RoadRunner supercomputer will be in use for quite some time forward. Also, IBM may reuse parts of the Cell architecture in some form [19].

In any case, Cell teaches us important general lessons about heterogeneous multi-core systems, and several important concepts of Cell programming will also be with us for the foreseeable future of power-efficient high-performance multicore programming: A combination and coordination of multiple levels of parallelism will be necessary. Multilevel memory hierarchies, whether explicit as in Cell or implicit as in cache-based systems, will require program optimizations for increased locality and overlapping of memory access latency with computation. Trade-offs between locality of memory accesses and load balancing have to be balanced. Explicit data transfer between local memory units over an on-chip network may become an issue even for future standard processor architectures, as cache-based SMP architectures may not scale up to very many cores. Instruction-level parallelism remains important but will fortunately be handled mostly by the compiler. Effective usage of SIMD operations is mandatory, and memory parallelism needs to be exploited. Finally, a certain amount of heterogeneity needs to be bridged. Managing this complexity while achieving high performance will require a joint effort by the programmer, language constructs, the compiler, the operating system, and other software tools.

8.8   Bibliographical Remarks

The IBM Cell SDK itself contains a programming tutorial [24], programming guides and manuals, and comprehensive documentation about Cell and its programming APIs. This material is also available on the IBM developerWorks web pages for Cell [28].

The recent book by Scarpino [50] is recommended reading for anyone trying to get familiar with programming Cell using the SDK. This book also contains detailed instructions for how to install Linux and the Cell SDK on (earlier versions of) a PlayStation 3. Appendix C gives an introduction to Cell programming with ALF.

A systematic method for selecting the optimal DMA multi-buffering scheme subject to local store space constraints has been proposed by Chen et al. [10].

Compiler techniques for Cell have been described, e.g., by Eichenberger et al. [14], Gschwind et al. [18], and Rosen et al. [48].

References to higher-level programming environments for Cell have been added in Section 8.5 as appropriate. Varbanescu et al. [56] give a survey and classification of programming environments for Cell.

Acknowledgments

The research of the author of this chapter was partly funded by EU FP7 (project PEPPHER, www.peppher.eu, grant 248481), by Vetenskapsrådet, SSF, Vinnova, CUGS, and Linköping University in Sweden.

The author thanks Duc Vianney from IBM, Ana Varbanescu from TU Delft, Jörg Keller from FernUniversität in Hagen, Dake Liu and his group at Linköping University, and Markus Schu from Micronas in München, for interesting discussions about Cell.

On-chip-pipelined mergesort for Cell was implemented by Rikard Hultén in a master thesis project at Linköping university. BlockLib was mainly developed by Markus Ålind and the NestStep runtime system for Cell by Daniel Johansson in earlier master thesis projects at Linköping University. We thank Erling Weibust, Carl Tengwall, Björn Sjökvist, Nils Smeds, and Niklas Dahl from IBM Sweden for commenting on this work and for letting us use their IBM QS-20 blade server. We thank Inge Gutheil and her colleagues at Jülich Supercomputing Centre for giving us access to their Cell cluster JUICE.

The author thanks Mattias Eriksson and Erik Hansson from Linköping university for discussions, for proof-reading and commenting on this chapter. Also, many thanks to George Russell and Uwe Dolinsky from Codeplay and to the anonymous reviewers for their comments.

However, any possibly remaining errors would be solely the author's own fault.

Disclaimers and Declarations

Trademarks

OpenCL is a trademark of Apple Inc.

RapidMind is a trademark of RapidMind Inc.

Still, we may have missed a few trademarked names, so there may indeed occur trademarks etc. that are not explicitly marked as such.

References

1. S.G. Akl. *Parallel Sorting Algorithms*. Academic Press, San Diego, CA, 1985.

2. M. Ålind, M. Eriksson, and C. Kessler. Blocklib: A skeleton library for Cell Broadband Engine. In *Proceedings of the ACM International Work*shop on Multicore *Software Engineering (IWMSE-2008) at ICSE-2008*, Leipzig, Germany, May 2008.

3. C. Augonnet, S. Thibault, and R. Namyst. Automatic calibration of performance models on heterogeneous multicore architectures. In *Proceedings of the International Euro-Par Workshops 2009, HPPC-2009, Springer LNCS 6043*, Delft, the Netherlands, pp. 56-65, 2010.

4. C. Augonnet, S. Thibault, R. Namyst, and M. Nijhuis. Exploiting the Cell/BE architecture with the StarPU unified rutime system. In *Proceedings of the Ninth International Workshop on Embedded Computer Systems: Architectures, Modeling, and Simulation (SAMOS'09), Springer LNCS 5657*, Samos, Greece, pp. 329–339, 2009.

5. E. Ayguade, R.M. Badia, D. Cabrera, A. Duran, M. Gonzales, F. Igual, D. Jimenez, J. Labarta, X. Martorell, R. Mayo, J. Perez, and E. Quintana-Orti. A proposal to extend the OpenMP tasking model for heterogeneous architectures. In *Proceedings of the Fifth International Workshop on Open MP: Evolving OpenMP in an Age of Extreme Parallelism IWOMP'09, Springer LNCS 5568*, Dresden, Germany, pp. 154–167, 2009

6. P. Bellens, J.M. Perez, R.M. Badia, and J. Labarta. CellSs: A programming model for the Cell BE architecture. In *Proceedings of the 2006 ACM/IEEE Supercomputing Conference (SC'06)*, Tampa, FL, November 2006.

7. B. Bouzas, R. Cooper, J. Greene, M. Pepe, and M.J. Prelle. Multicore framework: An API for programming heterogeneous multicore processors. Technical report, Mercury Computer Systems Inc., Chelmsford, MA, 2007.

8. S. Chellappa, F. Franchetti, and M. Püschel. FFT program generation for the Cell BE. In *Proceedings of the International Workshop on State-of-the-Art in Scientific and Parallel Computing (PARA)*, Trondheim, Norway, June 2008.

9. T. Chen, R. Raghavan, J.N. Dale, and E. Iwata. Cell broadband engine architecture and its first implementation—A performance view. *IBM J. Res. Dev.*, 51(5):559–572, September 2007.

10. T. Chen, Z. Sura, K. O'Brien, and K. O'Brien. Optimizing the use of static buffers for DMA on a CELL chip. In *Proceedings of the International Workshop on Languages and Compilers for Parallel Computers (LCPC'06)*, New Orleans, LA, Springer LNCS 4382, pp. 314–329, 2006.

11. P. Cooper, U. Dolinsky, A.F. Donaldson, A. Richards, C. Riley, and G. Russell. Offload—Automating code migration to heterogeneous multicore systems. In *Proceedings of the International HiPEAC-2010 Conference*, Pisa, Italy, Springer LNCS 5952, pp. 337–352, 2010.

12. U. Dolinsky, A. Richards, G. Russell, and C. Riley. Offloading parallel code on heterogeneous multicores: A case study using Intel threading building blocks on cell. In *Proceedings of the Many-Core and Reconfigurable Supercomputing Conference (MRSC-2010)*, Rome, Italy, www.mrsc2010.eu, March 2010.

13. A.F. Donaldson, U. Dolinsky, A. Richards, and G. Russell. Automatic offloading of C++ for the Cell BE processor: A case study using Offload. In *Proceedings of the International Workshop on Multi-Core Computing Systems (MuCoCoS'10)*, Krakow, Poland, pp. 901–906. IEEE Computer Society, February 2010.

14. A.E. Eichenberger, J.K. O'Brien, K.M. O'Brien, P. Wu, T. Chen, P.H. Oden, D.A. Prener, J.C. Shepherd, B. So, Z. Sura, A. Wang, T. Zhang, P. Zhao, M.K. Gschwind, R. Archambault, Y. Gao, and R. Koo. Using advanced compiler technology to exploit the performance of the Cell Broadband Engine™ architecture. *IBM Syst. J.*, 45(1), 2006.

15. T. El-Ghazawi, W. Carlson, T. Sterling, and K. Yelick. *UPC Distributed Shared Memory Programming*. Wiley-Interscience, Hoboken, NJ, 2005.

16. K. Fatahalian, T.J. Knight, M. Houston, M. Erez, D.R. Horn, L. Leem, J.Y. Park, M. Ren, A. Aiken, W.J. Dally, and P. Hanrahan. Sequoia: Programming the memory hierarchy. In *Proceedings of the 2006 ACM/IEEE Conference on Supercomputing*, ACM, Tampa, FL, 2006.

17. B. Gedik, R. Bordawekar, and P.S. Yu. Cellsort: High performance sorting on the cell processor. In *Proceedings of the 33rd International Conference on Very Large Data Bases*, Vienna, Austria, pp. 1286–1207, 2007.

18. M. Gschwind, D. Erb, S. Manning, and M. Nutter. An open source environment for Cell Broadband Engine system software. *Computer*, 40(6):37–47, 2007.

19. Heise online. SC09: IBM lässt Cell-Prozessor auslaufen, Hannover, Germany, http://www.heise.de/newsticker/meldung/SC09-IBM-laesst-Cell-Prozessor-auslaufen-864497.html, November 2009.

20. R. Hultén. Optimized on-chip software pipelining on the Cell BE processor. Master thesis, LIU-IDA/LITH-EX-A–10/015–SE, Linköping university, Sweden, 2010.

21. R. Hultén, C. Kessler, and J. Keller. Optimized on-chip pipelined mergesort on the Cell/B.E. In *Proceedings of the EuroPar-2010 Conference*, Ischia, Italy, Springer LNCS 6272, August 2010.

22. IBM. Basic linear algebra subprograms programmer's guide and API reference. In *SDK for Multicore Acceleration V3.0*, IBM Corporation, Armonk, NY, 2007.

23. IBM. IBM full-system simulator user's guide, V 3.0. *Software Development Kit for Multicore Acceleration V 3.0*, IBM Corporation, Armonk, NY, available at [28], 2007.

24. IBM. Programming tutorial DRAFT. *Software Development Kit for Multicore Acceleration V 3.0*, IBM Corporation, Armonk, NY, 2007.

25. IBM. SIMD Math library API reference manual. *SDK for Multicore Acceleration V3.0*, IBM Corporation, Armonk, NY, available at [28], 2007.

26. IBM. Accelerated Library Framework programmer's guide and API reference. *Software Development Kit for Multicore Acceleration V 3.1*, IBM Corporation, Armonk, NY, 2008.

27. IBM. *Cell Broadband Engine Programming Handbook, V 1.11*, IBM Corporation, Armonk, NY, Available at [28], May 2008.

28. IBM. Cell Broadband Engine. Document collection, https://www.ol.ibm.com/chips/techlib/techlib.nsf/products/Cell_Broadband_Engine (accessed August, 2011).

29. IBM. OpenCL development kit for Linux on Power. http://www.alphaworks.ibm.com/tech/opencl, 2009.

30. H. Inoue, T. Moriyama, H. Komatsu, and T. Nakatani. ΛΛ-sort: A new parallel sorting algorithm for multi-core SIMD processors. In *Proceeding 16th International Conference on Parallel Architecture and Compilation Techniques (PACT)*, Brasov, Romania, pp. 189–198. IEEE Computer Society, 2007.

31. J. JáJá. *An Introduction to Parallel Algorithms*. Addison Wesley, Reading, MA, 1992.

32. D. Johansson, M. Eriksson, and C. Kessler. Bulk-synchronous parallel computing on the CELL processor. In *Proceedings of the PARS'07: 21st PARS Workshop*, Hamburg, Germany, GI/ITG-Fachgruppe Parallel-Algorithmen, -Rechnerstrukturen und–Systemsoftware. In: *PARS-Mitteilungen* 24:90–99, ISSN 0177-0454, December 2007.

33. L.V. Kale and S. Krishnan. CHARM++: A portable concurrent object oriented system based on C++. In *Proceedings of the ACM SIGPLAN Conference on Object-Oriented Programming Systems, Languages and Applications (OOPSLA)*, Washington, DC, pp. 91–108, 1993.

34. J. Keller and C.W. Kessler. Optimized pipelined parallel merge sort on the Cell BE. In *Proceeding of the Second Workshop on Highly Parallel Processing on a Chip (HPPC-2008)*, Gran Canaria, Spain, August 2008. In E. Luque et al. *(Eds.): Euro-Par 2008 Workshops*, Springer LNCS 5415, pp. 127–136, 2009.

35. C. Kessler. Managing distributed shared arrays in a bulk-synchronous parallel environment. *Concurr. Comp. Pract. Exp.*, 16:133–153, 2004.

36. C.W. Kessler. NestStep: Nested parallelism and virtual shared memory for the BSP model. *J. Supercomput.*, 17:245–262, 2000.

37. C.W. Kessler and J. Keller. Optimized on-chip pipelining of memory-intensive computations on the Cell BE. In *Proceeding of the First Swedish Workshop on Multicore Computing (MCC-2008)*, Ronneby, Sweden. To appear in *ACM Computer Architecture News*, 36(5):36–45, 2009, 2008.

38. C.W. Kessler and J. Keller. Optimized mapping of pipelined task graphs on the Cell BE. In *Proceedings of the 14th International Workshop on Compilers for Parallel Computing (CPC-2009)*, Zürich, Switzerland, January 2009.

39. Khronos Group. The OpenCL specification, V 1.0. http://www.khronos.org/opencl, 2009.

40. M. Kistler, M. Perrone, and F. Petrini. Cell multiprocessor communication network: Built for speed. *IEEE Micro*, 26:10–23, 2006.

41. D. Kunzman. Charm++ on the Cell processor. Master's thesis, Department of Computer Science, University of Illinois, Urbama, IL, 2006. http://charm.cs.illinois.edu/newpapers/06-16/paper.pdf

42. T. Liu, H. Lin, T. Chen, J.K. O'Brien, and L. Shao. DBDB: Optimizing DMA transfer for the Cell BE architecture. In *Proceedings of the International Conference on Supercomputing*, New York, pp. 36–45, June 2009. ACM.

43. A. Lokhmotov, A. Mycroft, and A. Richards. Delayed side-effects ease multicore programming. In *Proceedings of the Euro-Par-2007 Conference*, Rennes, France, Springer LNCS 4641, pp. 641–650, 2007.

44. M. Monteyne. Rapidmind multi-core development platform. White paper, www.rapidmind.com, February 2008.

45. M. Ohara, H. Inoue, Y. Sohda, H. Komatsu, and T Nakatani. MPI Microtask for programming the CELL Broadband Engine^{TM} processor. *IBM Syst. J.*, 45(1):85–102, 2006.

46. I. Papp, N. Lukic, Z. Marceta, N. Teslic, and M. Schu. Real-time video quality assessment platform, In *Proceedings of the International Conference on Consumer Electronics 2009, ICCE '09*, Las Vegas, NV pp. 1–2, January 2009.

47. A. Richards. The Codeplay Sieve C++ parallel programming system. White paper, Codeplay Software Ltd., Edinburgh, UK, 2006.

48. I. Rosen, D. Edelsohn, B. Elliston, R. Eres, A. Modra, D. Nuzman, U. Weigand, and A. Zaks. Compiling effectively for Cell B.E. with GCC. In *Proceeding of the CPC-2009 International Workshop on Compilers for Parallel Computers*, Zürich, Switzerland, January 2009.

49. T. Saidani, C. Tadonki, L. Lacassagne, J. Falcou, and D. Etiemble. Algorithmic skeletons within an embedded domain specific language for the CELL processor. In *Proceedings of the 18th International on Par. Architectures and Compilation Techniques (PACT-2009)*, Novosibirsk, Russia, 2009. IEEE Computer Society.

50. M. Scarpino. *Programming the Cell Processor*. Prentice Hall, Upper Saddle River, NJ, 2008.

51. S. Schneider, J.-S. Yeom, B. Rose, J.C. Linford, A. Sandu, and D.S. Nikolopoulos. A comparison of programming models for multiprocessors with explicitly managed memory hierarchies. In *Proceedings of the ACM Symposium on Principles and Practice of Parallel Programming (PPoPP'09)*, Raleigh, NC, pp. 131–140, February 2009.

52. Sony Computer Entertainment Inc. PLAYSTATION® 3 worldwide hardware unit sales. http://www.scei.co.jp/ corporate/data/ bizdataps3_sale_e.html (March 2011), 2011.

53. A.J.C. van Gemund. The importance of synchronization structure in parallel program optimization. In *ICS '97: Proceedings of the 11th International Conference on Supercomputing*, Charleston, SC, pp. 164–171, 1997. New York: ACM.

54. H. Vandierendonck, S. Rul, M. Questier, and K.De Bosschere. Experiences with parallelizing a bio-informatics program on the Cell BE. In *Proceedings of the HiPEAC 2008, Springer LNCS 4917*, Goteberg, Sweden, pp. 161–175, January 2008.

55. A.L. Varbanescu, M. Nijhuis, A. Gonzalez-Escribano, H. Sips, H. Bos, and H. Bal. SP@CE: An SP-based programming model for consumer electronics streaming applications. In *Proceedings of the 19th International Workshop on Languages and Compilers for Parallel Computing (LCPC-2006), Springer LNCS 4382*, New Orleans, LA, pp. 33–48, 2007.

56. A.L. Varbanescu, H. Sips, X. Martorell, and R.M. Badia. Programming models for the Cell/B.E.: An overview with examples. In *Proceedings of the CPC-2009 International Workshop on Compilers for Parallel Computers*, Zürich, Switzerland, January 2009.

57. A.L. Varbanescu, H. Sips, K.A. Ross, Q. Liu, A. Natsev, J.R. Smith, and L.-K. Liu. Evaluating application mapping scenarios on the Cell/B.E. *Concurr. Comput. Pract. Exp.*, 21:85–100, 2009.

58. S. Williams, J. Shalf, L. Oliker, S. Kamil, P. Husbands, and K. Yelick. The potential of the Cell processor for scientific computing. In *CF'06: Proceedings of the Third Conference on Computing Frontiers*, Ischia, Italy, pp. 9–20, 2006. New York: ACM.

59. S. Williams, J. Shalf, L. Oliker, S. Kamil, P. Husbands, and K. Yelick. Scientific computing kernels on the Cell processor. *Int. J. Parallel Prog.*, 35(3):263–298, 2007.

60. D. Wu, Y.-H. Li, J. Eilert, and D. Liu. Real-time space-time adaptive processing on the STI CELL multiprocessor. In *Proceedings of the Fourth European Radar Conference*, Munich, Germany, October 2007.

61. D. Wu, B. Lim, J. Eilert, and D. Liu. Parallelization of high-performance video encoding on a single-chip multiprocessor. In *Proceedings of the IEEE International Conference on Signal Processing and Communications*, Dubai, UAE, November 2007.

Part IV

Emerging Technologies

Chapter 9

Automatic Extraction of Parallelism from Sequential Code

David I. August, Jialu Huang, Thomas B. Jablin, Hanjun Kim, Thomas R. Mason, Prakash Prabhu, Arun Raman, and Yun Zhang

Contents

9.1   Introduction

9.1.1   Background

Previous chapters have discussed many of the tools available to a programmer looking to parallelize code by hand. While valuable, each requires considerable effort by the programmer beyond that which is required for a typical single-threaded program. As more power is placed in the hands of the programmer, programmers must concern themselves with additional issues only posed by parallel programs—race conditions, deadlock, livelock, and more [1–3]. Furthermore, sequential legacy code presents its own problem. Not designed with parallelism in mind, the programmer may encounter great trouble in transforming the code to a parallel form.

Automatic parallelization offers another option to programmers interested in parallelizing their code. Automatic parallelism extraction techniques do not suffer the same limitations as manual parallelization. Indeed, through analysis, a compiler can often find parallelism in sequential code that is not obvious even to a skilled programmer. However, automatically parallelizing code poses a significant challenge of its own: finding means to extract and exploit parallelism in the code. The compiler must determine first what transformations can be done, and throughout this chapter, we will see a number of the possible techniques in the compiler's transformation toolbox. Simply put, this means the compiler must both find an exploitable region of code through analysis and transform the code into a parallel form. Analyses and transformations must work together to find and then exploit parallelization opportunities.

9.1.2   Techniques and Tools

There are many techniques for use in automatically parallelizing code. The names of some of those that will be discussed here are DOALL, DOACROSS, thread-level speculation (TLS), DOPIPE, and decoupled software pipelining (DSWP). All of these rely upon analysis and other tools to be successful. Data dependences can greatly degrade performance of parallelized code. As data dependences between load and store instructions create difficult scheduling problems on superscalar architectures, cross-thread dependences require communication and synchronization to maintain the strict sequential memory model on multicore architectures. Synchronization and communication are undesirable since they are relatively expensive operations and force faster threads to wait on slower threads. By careful scheduling, a parallelizing compiler can minimize the effect of data dependences. In fact, the quality of memory dependence analysis can strongly influences the performance of parallelized code. Unfortunately, determining whether or not a data dependence exists between two instructions is undecidable in general, and even with good analysis, many of these techniques are very brittle. A dependence that may rarely or even never manifest could greatly reduce opportunity for parallelization. By *speculating* that such a dependence does not exist, the compiler can greatly increase its opportunities. Speculation improves performance when the cost of misspeculation detection and the product of misspeculation recovery time and misspeculation frequency are less than the benefit of the speculation. Discussed in more detail later, alias, control, and value speculation have all been proposed to aid parallelization [4].

9.2 Dependence Analysis

9.2.1 Introduction

Dependence analysis is central to automatic parallelization. Although dependence analysis also has a variety of other applications like program slicing for visual tools [5], scalar compiler optimizations [6], and program verification [7], automatic parallelization is one of the most important clients of a robust dependence analysis. The accuracy and efficiency of dependence analysis can directly affect various important aspects of automatic parallelization including scope of applicability, performance scalability of the parallelized program, and compilation time.

So, what is dependence analysis? Simply put, it is a broad term used to describe compiler analyses to disprove various kinds of dependences that may exist between different imperative program units. These units may be low-level assembly instructions, individual statements in different compiler intermediate representations, or even high-level language program statements or functions. Quite often, the abstraction level at which dependence analysis is done is determined by the kind of optimization for which it is intended to be used. For ease of understanding, we will assume that the analysis is done on high-level language statements.

The example code snippet shown in Figure 9.1 is used to explain the concept of a dependence and dependence analysis and show its importance in automatic parallelization. The program traverses a linked list, calculates a value from each node, and then adds it to the total cost of the list. Statement 5, the blank statement, will be useful for some of our examples, but simply represents further code.

Before examining code for dependences, we must understand why we are doing so. For instance, if there are two loops in a program, we might be interested to find out whether the two loops could execute at the same time, in parallel within separate threads of control. This would be possible if the loops were not reading and writing to the same variable (such a scenario would result in a *race condition* where the two threads race to read/write a shared variable and the output is dependent upon the nondeterministic timing of the operations). However, since we have only one loop, our first question is, "is it possible to execute successive iterations of the loop in

```
1:    while (node)
2:       {ncost  =  calc(node);
3:       cost  +=  ncost;
4:       node  =  node->next;}
5:       ...
```

FIGURE 9.1:   Example to illustrate data dependence analysis.

parallel?" The answer here is, at first glance, categorically no—each iteration depends on knowing the node operated on in the previous iteration so that it can advance to the next node. A parallelizing compiler would infer this property by formalizing the concept of a *data dependence*.

When we further consider the code from Figure 9.1, we notice another type of dependence. The loop will only continue executing so long as the node is not NULL. Since the execution of the loop body (statements 2, 3, 4) is dependent on whether the condition tested in the while statement is true, these three statements are said to be *control dependent* on statement 1.

Sections 9.2.2 and 9.2.3 formalize the concepts of data dependence and control dependence. Section 9.2.4 then describes a program structure that represents both kinds of dependences in a uniform way that is useful for parallelization.

9.2.2   Data Dependence Analysis

9.2.2.1   Data Dependence Graph

In order to analyze dependences, it is convenient to visualize a program as a directed graph, called the *control flow graph*, whose nodes represent the program statements and whose edges represent control flow relations determined by sequential composition and branch (conditional/unconditional) statements. The control flow graph of the program shown in Figure 9.1 is shown in Figure 9.2.

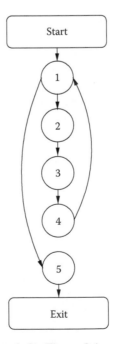

FIGURE 9.2:   Control flow graph for Figure 9.1.

A data dependence exists between two statements A and B if the following conditions hold:

- Both A and B either write or read the same program variable, or in general, the same memory location.

- At least one of A or B does a write to the common memory location.

- There is a valid control flow path in the program from A to B or from B to A.

The data dependence relation between statements of a program can be represented by means of a graph called the *data dependence graph*. The data dependence graph (DDG) is defined by $G = (N, E)$ where N, the node set, has one node for every statement in the function and E, the edge set, has one edge for every pair of nodes that represent statements with a data dependence between them.

Three kinds of data dependence edges can be distinguished:

1. A *flow dependence edge* from node A to B, whenever A writes to the same memory location that is read by B and there is a valid control flow path from A to B. This is also called a "read-after-write" (RAW) dependence.

2. An *anti-dependence edge* from node A to B, whenever A reads from the same memory location that is written to by B and there is a valid control flow path from A to B. This is also called a "write-after-read" (WAR) dependence.

3. An *output dependence edge* from node A to B, whenever A writes to the same memory location that is written to by B and there is a valid control flow path from A to B. This is also called a "write-after-write" (WAW) dependence.

The data dependence graph for the program in Figure 9.1 is shown in Figure 9.3. The nodes are the same as the nodes in the control flow graph of Figure 9.2 (statement 5 is presumed to have no data dependences for our purposes and thus omitted), while the edges represent the data dependence relation and are of three kinds: f edges representing flow dependence, a edges representing anti dependence, and o edges representing output dependence.

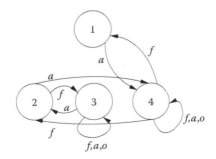

FIGURE 9.3: Data dependence graph for Figure 9.1.

9.2.2.2 Analysis

In this section, we give an overview of the various analyses that have been proposed by compiler researchers to efficiently construct the data dependence graph with varying degrees of accuracy. All analyses discussed in this section were proposed in the past few decades. Traditionally, there has always been a trade-off between the scalability of a data dependence analysis and the accuracy of the constructed data dependence relation. First, it should be noted that determining the exact dynamic data dependence relation between all statements in a program is undecidable in general [8]. This means any data dependence relation that is constructed by an analysis is a *safe approximation* of the actual dependences manifested at runtime. The running time and memory requirements of a data dependence analysis are, therefore, inversely related to the accuracy of the analysis: a fast analysis usually reports more dependences than a slower, more accurate analysis [9].

The complexity of a data dependence analysis also depends on the properties of the language being analyzed: usually it is easier to analyze data dependences in languages like FORTRAN 77 that permit only restricted uses of pointer variables than to analyze languages like C where the use of pointers and heap allocated data structures create complicated data dependence relations through multiple levels of indirection. Languages like Java make the analysis of pointers simpler with features like strong typing and the absence of pointer arithmetic but still have other features like polymorphism, and dynamic class loading, which introduce other complications that a data dependence analysis has to contend with. In the following paragraphs, we will discuss the different types of analyses in increasing order of accuracy (and hence increasing order of complexity).

The *reaching definitions analysis* [10] is a simple data-flow analysis that can be used to determine flow dependences between statements in a program that do not have pointers (or in case of a low-level intermediate representation, the flow dependences between register/temporary variables).

Relatedly, *liveness analysis* [10] is a data flow analysis that is used to determine at each program point the variables that may be potentially read before their next write. Applying this definition in the context of loops, a variable v is said to be a *live-in* into a loop if v is defined at a program point outside the loop and used at a point inside the loop and there exists a path from the point of definition to the point of use with no intervening definitions in between. Similarly, a variable is said to be a *live-out* of a loop if it is defined at a point inside the loop and used outside the loop and there exists a path from the point of definition to the point of use with no intervening definitions in between.

Reaching definitions analysis along with liveness analysis, can be used for determining flow and anti-dependences in a data dependence graph. Output dependences for programs with no pointer aliasing (two pointers are aliases of each other if they may point to the same memory location at runtime) can be determined by intersecting the variables defined at program points that are reachable from each other. In programs written in languages that permit the use of pointer variables, *pointer analysis* is employed for determining the data dependences between statements. Given a pair of pointer variables, a pointer analysis determines whether the two variables can point to the same memory location at runtime (may alias or may not alias). Pointer analysis plays an important role in disproving data dependences in pointer-based languages like C and Java, as can be inferred by the large amount of research literature describing its various forms [11]. Pointer analysis is further augmented by *shape analysis* that tries to determine the form or "shape" of allocated data structures, disambiguating trees from linked lists, etc. For more details about flow-sensitive analysis [6], field-sensitive analysis [12] and context-sensitive analysis [13], the interested reader is referred to the relevant citations. More detailed discussion of shape analysis and the logic used to describe data structures is also beyond the scope of this book [14–16].

Both pointer and shape analysis, in their basic form, are not loop and array-element sensitive; they do not disprove dependencies between statements accessing different array elements that execute on different iterations of a loop (called inter-iteration or loop-carried dependences). The main reason for this is that the complexity of maintaining and propagating distinct points to sets for each array element on different iterations of a loop is prohibitive in a pointer-based language like C. Inter-iteration dependences are often those with which we are most concerned for our parallelizations. Furthermore, the prevalence of array-based loops in scientific programs has led researchers to develop a separate class of dependence analysis called the *array dependence analysis* [17–19]. Array dependence analysis expresses the array accesses in a loop in the form of linear functions of the loop induction variables and then goes on to express the conditions for the existence of dependences between these accesses as an integer linear programming problem does not admit a feasible solution, then it proves the absence of a dependence between the multiple iterations of a loop and hence they can be executed in parallel. Although the solving of ILP in the general case is NP-hard, there have been different kinds of tests like the GCD test and Banerjee test [17,20] that have been successfully applied for specific commonly occurring cases. It should be noted that array dependence analysis is easier to apply in languages like FORTRAN 77 (which is traditionally the language of choice for scientific computation) which have restrictive pointer aliasing than in languages like C and Java where the use of dynamically allocated arrays and the use of pointer indirection requires both pointer analysis and array dependence analysis in a single framework to be effective [21].

9.2.3 Control Dependence Analysis

9.2.3.1 Control Dependence Graph

A control dependence between two statements A and B can be defined using the concept of post-dominance on the control flow graph. A node X in the control flow graph is said to *post-dominate* a node Y in the graph if every path from node Y to an exit node of the graph has to necessarily pass through X. For example, in control flow graph Figure 9.2, node 5 post-dominates 1 but 4 does not post-dominate 1. Using the

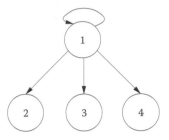

FIGURE 9.4:   Control dependence graph for Figure 9.1.

post-dominance relation, control dependences can easily be computed. A node B is said to be control dependent on node A if:

- There exists a valid execution path p in the control flow graph from A to B

- B post-dominates every node C in p

- B does not post-dominate A

A *control dependence graph* is a graph G = (N, E) where N is the set of nodes in the graph and E is the set of edges in the graph. There is one node representing every statement in the program and an edge is created from A to B whenever B is control dependent on A. The control dependence graph for the control flow graph of Figure 9.2 is shown in Figure 9.4.

9.2.3.2   Analysis

Ferrante et al. [22] give an algorithm for computing the control dependence edges using the idea of dominance frontiers on the reverse control flow graph of a program. In order to understand the term "dominance frontier," we need to first define what we mean by the term "dominance." A node A is said to *dominate* a node B in a control flow graph if every path from start node of the control flow graph to B has to necessarily pass through A. It is not difficult to see that dominance is transitive, which means that it is possible to arrange the nodes of a control flow graph in the form of a *dominator tree*, where there is an edge from A to B (nodes are the nodes of a control flow graph) if A is an *immediate dominator* of B (i.e., there is no other node C such that A dominates C and C dominates B). For example, the dominator tree for the control flow graph of Figure 9.1 is shown in Figure 9.5. The *dominance frontier* of a node A in the control flow graph is defined as the set of all nodes X such that A dominates a predecessor of X but not X itself. The *iterated dominance frontier* of node A is the transitive closure of the dominance frontier relation for the node A. The reverse control flow graph is the control flow graph of a program with its edges reversed. Ferrante et al.'s algorithm constructs the dominance frontiers of every node in the reverse control flow graph, which is the same as the *post-dominance frontier* relation. For every node B in the post-dominance frontier of a node A, an edge from B to A is created in the control dependence graph. For more details on the efficient

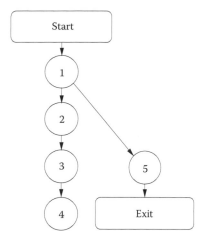

FIGURE 9.5: Dominator tree for Figure 9.1.

construction of control dependences, we refer the reader to the paper by Ferrante et al. [22].

9.2.4 Program Dependence Graph

The program dependence graph (PDG) [22] is a single compiler intermediate representation that represents both the data dependence and control dependence relation in a uniform manner. The *program dependence graph* of a program is a graph $G = (N, E)$, where N is the set of nodes and E is the set of edges in the graph. There is one node for every statement in the program. There are two kinds of edges: *control dependence edges* and *data dependence edges*. The program dependence graph is therefore the union of the data dependence graph and control dependence graph of a program. For example, the program dependence graph corresponding to the program in Figure 9.1 is shown in Figure 9.6.

9.3 DOALL Parallelization

9.3.1 Introduction

DOALL parallelization is one of the simplest ways to parallelize a loop. The technique schedules each iteration of a loop for execution in parallel without any communication between the iterations. In addition to the base transformation, this section describes privatization, reduction, speculative DOALL, and other transformations that enable DOALL on a wide range of applications. These techniques are highly effective in parallelizing code, which often exhibits the sorts of dependence patterns that prevent DOALL parallelization.

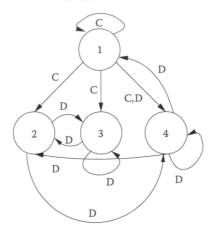

FIGURE 9.6:   Program dependence graph for Figure 9.1. The data and control dependence edges are marked D and C, respectively.

In DOALL parallelization, all iterations are executed simultaneously without any synchronization. Therefore results can be wrong if any dependence edge exists between iterations. Bernstein [23] described three conditions for two program blocks i and j to be safely executed in parallel.

1. ReadSet(i) $\cap$ WriteSet(j) $= \oslash$

2. WriteSet(i) $\cap$ ReadSet(j) $= \oslash$

3. WriteSet(i) $\cap$ WriteSet(j) $= \oslash$

If a loop iteration does not have any other loop exit condition except one that depends on the loop induction variable, it can be considered as a program block, and Bernstein's conditions can be applied at the iteration level. The conditions can be translated into the following dependence conditions that can be checked by a compiler in dependence analysis. If a loop satisfies all conditions, the loop can be parallelized in a DOALL manner.

1. No loop-carried WAR dependences

2. No loop-carried RAW dependences

3. No loop-carried WAW dependences

4. No loop-carried control dependences except the loop induction–based exit condition

Examples in Figure 9.7 cannot be parallelized using DOALL because they do not satisfy the DOALL dependence conditions. Figure 9.7a violates the no loop-carried WAR dependence condition because a[i − 1] is read in an earlier iteration than when it is written. If the loop is parallelized using DOALL, an updated a[i] can be read when iteration i + 1 is executed before iteration i.

```
int  a[N],  b[N];
for(i=1;i<N;i++){
   a[i-1]  =  a[i]+b[i];
}
(a)
```

```
int  a[N],  b[N];
for(i=0;i<N-1;i++){
   a[i+1]  =  a[i]+b[N];
}
(b)
```

```
int  a[N],  b[N];
int  c;
for(i=0;i<N;i++){
   c  =  a[i]+b[i];
}
(c)
```

```
int  a[N],  b[N],  c[N];
for(i=0;i<N;i++){
   c[i]  =  a[i]+b[i];
   if(a[i]>0)  break;
}
(d)
```

FIGURE 9.7: Bad DOALL candidates. (a) WAR dependence. (b) RAW dependence. (c) WAW dependence. (d) Loop exit condition.

Figure 9.7b violates the no loop-carried RAW dependence condition because a[i] is written in the previous iteration. Like the previous case, if iteration i+1 is executed before iteration i the old a[i + 1] value can be read.

Figure 9.7c violates the no loop-carried WAW dependence condition because two different iterations write their results to the same location c. If this loop is parallelized using DOALL and c is read after the loop, the wrong value can be read.

Figure 9.7d violates the loop exit condition. If the loop is parallelized using DOALL, later iterations may be executed even though the loop should have exited before iteration N.

9.3.2 Code Generation

The example in Figure 9.8a can be safely parallelized in a DOALL manner because it satisfies all DOALL dependence conditions. To parallelize this loop using M threads, first, a unique ID from 0 to $M - 1$ is assigned to each thread. This ID will be used for a thread to select iterations to execute in the parallelized code. Second, the number of iterations is divided into M groups. The iterations can be grouped based either on modular value or range. In a modular value–based scheme, thread i executes iteration j where $i = j$ MOD N. In a range-based scheme, thread i executes iteration j where $j \in [i{\times}N/M, (i + 1){\times}N/M)$. In general, range-based grouping shows better performance than a modular-based grouping because it often has better cache locality. The code in Figure 9.8b shows the parallelized code with range-based grouping, and Figure 9.9 shows its execution model.

Even though DOALL can be applied to the example, the shown dependence conditions are too strict for DOALL to be generally applicable. Fortunately, some dependences can be ignored due to special conditions or by using certain transformations.

In Figure 9.10a, there are loop-carried WAW and WAR dependences on tmp so it cannot be parallelized using DOALL because the wrong tmp value can be read after for loop execution and tmp updated in a later iteration can be read. If tmp is not read after the for loop, the loop-carried WAW dependence can be ignored. Liveness

```
                              Sequential  program  variables:
                              int  a[N],  b[N],  c[N],  tid[M];

int  a[N],  b[N],  c[N];      Parallel  code:
for(i=0;i<N;i++){                 void  doall(int  *tid){
  c[i]  =  a[i]+b[i];               int  id  =  *tid;
}                                   for(i=id*N/M;i<(id+1)*N/M;i++){
                                      c[i]  =  a[i]+b[i];
                                    }
                                  }
  (a)                             (b)
```

FIGURE 9.8: DOALL example. (a) Original program. (b) Parallelized program.

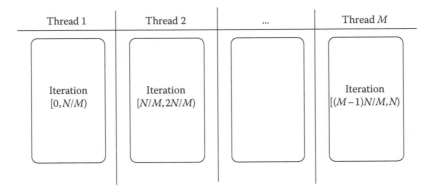

FIGURE 9.9: DOALL execution model.

analysis can tell whether a variable will be used later in the program execution. If `tmp` is not live-out, the WAW dependence can be ignored. The WAR dependence can be also ignored if `tmp` is privatized. Therefore, using liveness analysis and privatization, this loop can be parallelized using DOALL as shown in Figure 9.10b. Note that since `tmp` is a local variable in each thread, `tmp` is privatized.

In all of these cases, the intra-iteration dependences are inevitably respected as any single iteration is executed on a single thread, and thus the compiler does not need to handle these dependences.

9.3.3 Advanced Topic: Reduction

Summing the elements of an array is an operation that is common across many programs. However, since there is a loop-carried dependence on the sum variable, the loop cannot be parallelized by the DOALL technique directly. In this case, we can use what is called a sum reduction to parallelize the operation. Reduction can be applied not only to summing elements, but to all associative and commutative operations, including multiplication, finding the maximum or minimum element in

```
                               Sequential   program   variables:
                               int   a[N],   b[N],   c[N],   d[N],   tid[M];

int   a[N],   b[N],   c[N];
int   tmp;                     Parallel   code:
for(i=0;i<N;i++){                 void   doall(int   tid){
   tmp   =   a[i]+b[i];              int   id   =   tid;
   d[i]   =   tmp*c[i];             int   tmp;
}                                    for(i=id*N/M;i<(id+1)*N/M;i++){
                                        tmp   =   a[i]+b[i];
                                        d[i]   =   tmp*c[i];
                                     }
                                  }
(a)                            (b)
```

FIGURE 9.10: DOALL example with WAW and WAR dependences. (a) Original program. (b) Parallelized program.

```
                                    int   a[N],   b[N],   s[M];
                                    int   sum   =   0;
                                    for(j=0;j<M;j++){
int   a[N],   b[N];                    s[j]   =   0;
int   sum   =   0;                  }
for(i=0;i<N;i++){              L1:   for(j=0;j<M;j++){
   sum   =   sum   +   a[i]*b[i];  L2:      for(i=j*N/M;i<(j+1)*N/M;i++){
}                                        s[j]   =   s[j]   +   a[i]*b[i];
printf("sum:   %d\n",   sum);          }
(a)                                 }
                                    for(j=0;j<M;j++){
                                       sum   =   sum   +   s[j];
                                    }
                              (b)   printf("sum:   %d\n",   sum);
```

FIGURE 9.11: Reduction example. (a) Original program. (b) Parallelized program.

an array and counting the number of elements meeting some condition. These latter two reductions are called as max/min reduction and count reduction [24].

We will take a dot product code in Figure 9.11a as an example for sum reduction. As mentioned earlier, this loop cannot be parallelized with the DOALL technique because there are loop-carried WAW and WAR dependences on sum. Add operations are commutative and associative, and so we can actually perform any set of add operations in any order. What this means is that if the sum operations are divided into several groups and then accumulate all partial-sums from the groups after computation, the correct result will be achieved. Therefore, this example can be changed into code shown in Figure 9.11b.

Partial-sums are initialized in the first loop, and they are calculated in each group in the second loop. In the last loop, the final sum is calculated by accumulating sub-sum results. Unlike the previous code, in this code, L1 can be parallelized using DOALL because L2 loops for different j can be executed at the same time.

9.3.4 Advanced Topic: Speculative DOALL

As we have seen so far, DOALL parallelization can be applied to a wide range of loops with the aid of some enabling transformations. However, it is common to insert error-checking code in an otherwise DOALL parallelizable loop, as in the following example (Figure 9.12a).

The loop in Figure 9.12a is almost exactly the same as in Section 9.3.2 except that there is one conditional branch for error checking. This early-exit branch adds inter-iteration dependence edges that prevent DOALL parallelization because we do not know whether the next iteration is going to be executed or not depending on the value of a[i] and b[i] in this iteration, though the early branch may never happen at runtime.

The solution is to speculate this rarely occurring branch [25]. To enable DOALL parallelization using speculation, a compiler first detects which dependences should be speculated. In the example, the control dependences created by statement 2 happen atmost once per invocation, so they are good candidates for speculation. Speculating that this branch will not occur, the compiler must then insert code for detecting

```
      int   a[N],   b[N],   c[N];
1:    for(i=0;i<N;i++){
2:        if   (a[i]<0   ||   b[i]<0)   break;
3:        c[i]   =   a[i]   +   b[i];
      }
```

(a)

```
Thread  1:                        Thread  2:

                                  for   (i   =   N/2   ;   i   <N;   i++)   {
misspec   =   0;                      if   (misspec   ==   1){
for   (i   =   0;   i<   N/2;   i++}   {       rollback();
    if   (a[i]<0   ||   b[i]<0)   {          break;
    misspec   =   1;                     }
    break;                            if   (a[i]<0   ||   b[i]<0)
    }                                     break;
    c[i]   =   a[i]   +   b   [i];        c[i]   =   a[i]   +   b[i];
}                                     }
if   (!misspec)                   send(thread1);
    receive(thread2);
```

(b)

FIGURE 9.12: Speculative DOALL example. (a) Original program. (b) Parallelized program.

misspeculation into the loop to guarantee correctness. Misspeculation recovery functionality is needed to recover if misspeculation occurs. This misspeculation detection and recovery can be done either with hardware support, using something such as transactional memory, or in software, with software transactional memory. The compiler can transform the original loop into the one in Figure 9.12b. All loop iterations can now be executed speculatively in parallel.

Now, as before, we can spawn more threads to run the loop speculatively. If the early exit condition is never satisfied, the loop will execute in parallel with no extra cost. If there is an error in the input data and the loop does exit early, we will need to recover from the misspeculation. This involves discarding all computation done in later iterations (in our example, those done in thread 2) and restoring the values originally in array c after the misspeculation takes place. There is a trade-off between enabling more DOALL parallelism and the cost of misspeculation in real-world applications. A compiler can decide when and where to speculate a value or branch to remove data or control dependences in the PDG at compile time.

In real-world programs, such as scientific applications, vector calculations are very common and, in fact, often dominate. While many array and vector computations can be parallelized in a DOALL fashion, it is also common to have bounds or threshold-checking conditional branches as we saw in our simple example. As a practical example, in a support vector machine program, the goal is to compute a division plan that divides the dataset best. The computation loop finishes when the result is good enough. That is when a certain threshold is reached. This rarely taken early-exit branch inhibits a large amount of otherwise parallelizable code. With the speculative DOALL technique, we can get the parallelism back with only one misspeculation.

9.3.5   Advanced Topic: Further Techniques and Transformations

While the DOALL technique and the extensions we have seen through reduction and speculation are frequently useful in parallelization of scientific code (and occasionally more general purpose programs), their utility is still severely limited by the structure of the program constructed by the programmer. The structure may inhibit parallelism entirely or it may simply make it unprofitable to parallelize in its current form because the parallelized code must communicate and synchronize too frequently. Fortunately, we have at our disposal a great number of ways to transform the code in order to discover parallelism. We will discuss a few possible transformations here.

One technique, perhaps the simplest technique, that can be applied to make parallel execution more profitable is *loop interchange* [26]. As an example, let us look at the code in Figure 9.13a. We can see fairly clearly, applying concepts from earlier in this section, that the loop at statement 2 is immediately parallelizable by the DOALL technique. In fact, ignoring the i indices that would remain constant across an invocation of the loop, this is precisely the same as the loop seen at the beginning of Section 9.3.2. However, in its current form, there is a dependence across iterations of the outer loop (statement 1), which prohibits us from parallelizing the outer loop and reducing communication.

```
1:   for  (i  =  0;  i  <  N;  i++)  {
2:      for  (j  =  0;  j  <  M;  j++)  {
3:         A[i+1][j]  =  A[i][j]  +  B[i][j];
4:      }
5:   }
```
              (a)

```
1:   for  (j  =  0;  j  <  M;  j++)  {
2:      for  (i  =  0;  i  <  N;  i++)  {
3:         A[i+1][j]  =  A[i][j]  +  B[i][j];
4:      }
5:   }
```
              (b)

FIGURE 9.13:   Loop interchange. (a) Original loop. (b) Interchanged loop.

The loops, however, can be interchanged. That is, their order can be swapped without changing the resultant array. If we make this interchange, we generate the code seen in Figure 9.13b. In this transformed code, our dependence recurrence is now across the inner loop while iterations of the outer loop fulfill our criteria for making a successful DOALL parallelization. The benefit is that we no longer need to synchronize and communicate results at the end of every invocation of the inner loop. Instead, we can distribute iterations of the outer loop to different threads, thus completing each inner loop invocation on a different thread, and need only communicate results once at the end.

Another useful loop transformation that can obtain DOALL parallelism in scientific programs where none exists in its initial form is *loop distribution*. Loop distribution takes what is originally a single loop with multiple statements with inter-iteration dependences and splits it into multiple loops where there are no inter-iteration dependences. Let us take the code in Figure 9.14a as an example. As we can see, there is an inter-iteration flow dependence on the array *A*. We cannot immediately use a DOALL parallelization on our loop because of this dependence. However, we note that all of the computations in statement 2 could be completed independently of the calculations in statement 3 (statement 2 does not depend on statement 3). Statement 3, in fact, could wait and be executed after all calculations from statement 2 are executed. The compiler, recognizing that statement 3 depends on statement 2 but that statement 2 has no loop-carried dependences, can distribute this loop.


```
1:   for(i  =  0;  i  <  N;  i++)  {      1:   for(i  =  0;  i  <  N;  i++)
2:      A[i]  =  A[i]  +  C[i];            2:      A[i]  =  A[i]  +  C[i];
3:      B[i]  =  A[i  -  1]  +  10;        3:   for(i  =  0;  i  <  N;  i++)
4:   }                                     4:      B[i]  =  A[i  -  1]  +  10;
```
              (a)   (b)

FIGURE 9.14:   Loop distribution. (a) Original loop. (b) Distributed loops.

The code generated by distributing this loop is shown in Figure 9.14b. If we look at this code, the statements in the first loop (statements 1 and 2) very closely resemble our first DOALL example from Section 9.3.2; it is absolutely a good candidate for the DOALL technique. The second loop now depends on the first loop but has no loop-carried dependences. It, too, may be further transformed by the DOALL technique. By small changes to the code structure, the compiler is able to take one sequential loop and transform it into two perfect candidates for the DOALL technique.

A more detailed discussion of both loop interchange and loop distribution can be found in various modern compiler texts such as the text by Allen and Kennedy [24].

While these two straightforward transformations are good examples of code transformations that enable extraction of more parallelism with the DOALL technique in scientific code, they are only samples of the vast toolbox of transformations that enable further DOALL parallelism. We have seen a few other important transformations and techniques with privatization, reduction, and speculative DOALL. A few others are worthy of mention here but cannot be covered in detail.

We have only covered a few techniques amongst many for creating parallelism from non-parallel loops. Others include *loop reversal* and *loop skewing*. While the loop transformation techniques discussed here unlock parallelism in well-nested loops, imperfectly nested loops require other techniques to find parallelism. For this, we may turn to *multilevel loop fusion* [27]. While these methods all unlock opportunities for parallelism, we are generally at least as concerned with how we actually schedule and distribute the workload. A technique called *strip mining* is useful in vectorization techniques for packaging parallelism in the best manner for the hardware [28].

Finally, there is one technique that brings together many of these transformations and techniques, including loop fusion and loop distribution amongst others, in order to find parallelism. This transformation, called an *affine transform*, partitions code into what are called *affine partitions*. Affine transforms are capable not only of generating parallelism that fits with the DOALL technique but can find parallelism in array-based programs with any number of communications [29]. In fact, it can be used to find pipeline parallelism as in DOPIPE, which will be discussed in detail in Section 9.5. This makes DOALL among the most powerful transformations for extracting parallelism from scientific code.

9.4   DOACROSS Parallelization

9.4.1   Introduction

In the previous section, we introduced DOALL parallelization. DOALL scales very well with the number of iterations in a loop; however, its applicability is limited by the existence of loop-carried dependences. In this section, we will introduce DOACROSS parallelization that is able to exploit parallelism in loops with cross iteration dependences. In advanced topics, we will describe how speculation is employed in DOACROSS to enable more parallelism. Thread-level speculation (TLS) will be

introduced as a widely recognized example, and the TLS technique known as speculative parallel iteration chunk execution (Spice) will be discussed.

DOACROSS schedules each iteration on a processor with synchronization added to
enforce cross-iteration dependences. In Section 9.2, we saw in Figure 9.1 a loop that
contained loop-carried dependences. The corresponding program dependence graph
was formed in Figure 9.6. If we assign different iterations of the loop onto different
cores, synchronizations are communicated between different cores. As we saw with
DOALL (Section 9.3), since a single iteration is executed on a single thread, we will
inevitably respect the intra-thread anti-dependences as ignored in Figure 9.15a shows
the DOACROSS execution model for this loop. As one iteration is executed on a
single thread, the intra-iteration dependences are ignored in the figure.

In Figure 9.15a, we assume that the inter-core communication latency is only 1
cycle and we are able to achieve a speedup of 2× over single-threaded execution.
However, since synchronizations between different cores are put on the critical path,
DOACRORSS is quite sensitive to the communication latency. If we increase the
communication latency to 2 cycles, as seen in Figure 9.15b, it now takes DOACROSS

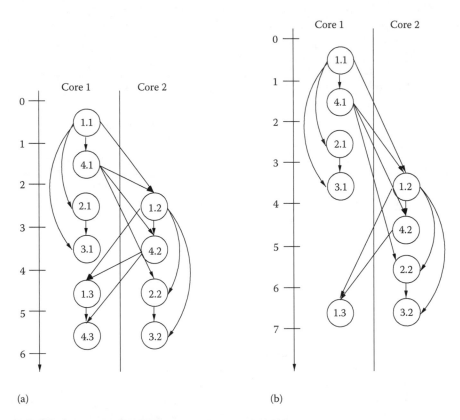

FIGURE 9.15:   DOACROSS example. (a) DOACROSS with latency = 1 cycle. (b)
DOACROSS with latency = 2 cycles.

3 cycles instead of 2 to finish a certain iteration. So, inefficient implementation of inter-processor communication can greatly reduce the benefit of parallel execution.

9.4.2 Code Generation

To do a DOACROSS transformation, the loop-carried dependence graph is constructed first and then the synchronization instructions are inserted into the loop to enforce the dependences.

Different synchronization schemes have been proposed in previous work. Some work has proposed compiler algorithms that insert special synchronization instructions into DOACROSS loops [30]. Another example computes the access orders of each data element at compile time and then inserts data synchronization instructions to enforce those dependences [31]. Other research has taken aim at distributed shared-memory multiprocessors [32]. It uses direct communication and a static message passing method. It proposes compiler algorithms to generate communication primitives for DOACROSS loops. The primitives use the nearest shared memory (NeSM) as communication buffers to eliminate major inter-processor communication overhead.

DOACROSS code generation depends on the synchronization scheme. Here, we take the algorithm proposed by Midkiff and Padua [30] as an example. In the loop shown in Figure 9.16.

The only cross-iteration dependence exists from statement 2 to statement 1. Before statement 1 in iteration i can execute, it needs to make sure that statement 2 in iteration $i-2$ has already finished execution.

To enforce the dependence, synchronization instructions will be inserted into the loop. Here we use synchronization instructions test and testset. The semantics of these two instructions are

- test (R, Δ): R is the variable to be tested and Δ is the dependence distance. R is shared by all of the threads and it is always equal to the number of iterations that have finished. The test instruction does not complete execution until the value of the variable is at least equal to the number of the iteration containing the source of the dependence being synchronized. In other words, test (R, Δ) in iteration i will keep executing until value of R becomes at least $i - \Delta$.

- testset (R): testset checks whether the testset in the previous iteration has finished or not. If it has, then testset updates the R value to the present iteration number.

After transformation, the original loop becomes the one in Figure 9.17.

```
        for (i=0; i<N; i++) {
1:        A[i] = B[i-2] + C[i];
2:        B[i] = A[i] + D[i];
        }
```

FIGURE 9.16: Original program.

```
doacross  I  =  1,  N
    test(R,  2)
1:  A[I]  =  B[I-2]  +  C[I];
2:  B[I]  =  A[I]  +  D[I];
    testset(R)
end  doacross
```

FIGURE 9.17:   Loop after DOACROSS transformation.

Here, test (R, 2) makes sure that before statement 1 can execute in the present iteration i, iteration 1, 2 has already finished executing because only after that will the value R be set to the iteration number by instruction testset (R). Now different iterations can be run in parallel on different cores.

9.4.3   Advanced Topic: Speculation

The algorithms introduced earlier depend on the accurate detection of dependence distances across iterations. However, for more general-purpose programs, dependences can be ambiguous. To reduce the communication amount between processors, speculation is necessary.

In a DOACROSS-style parallelization, we speculatively execute iterations in parallel as if there are no loop-carried dependences. This is called Thread-level speculation (TLS). There are two common speculations:

1. Memory alias speculation: This assumes that loads in later iterations do not conflict with stores in the earlier iterations. If the speculation turns out to be false due to dependences between the iterations, the speculatively executed iterations are squashed and restarted. As long as the conflict between loads and stores in different iterations are infrequent, performance does not suffer. If the dependences between loop iterations manifest frequently, alias speculation suffers from high misspeculation rates, causing a slowdown when compared with single-threaded execution. This can be overcome by synchronizing those store-load pairs that conflict frequently.

2. Value prediction: This uses value predictors to predict the live-ins for future iterations and speculatively executes the future iterations with these predicted values. Different value prediction techniques, such as last value predictors, stride predictors, and trace-based predictors have been proposed. Some common uses of value speculation are biased branch speculation (which speculates that a condition for a branch is unlikely to be met), load prediction (which predicts what value will be loaded), and silent store speculation (which speculates that the store will not actually change the existing value).

Another type of speculation called memory value speculation combines elements of both of the aforementioned speculation types by speculating upon the value read from memory at a certain point, without specific regard to the memory location from which it is read.

```
1:   while(node)   {
2:      ncost   =   calc(node);
3:      cost   +=   ncost;
4:      if   (cost   >   T)
5:         break;
6:      node   =   node->next;
7:   }
```

FIGURE 9.18:   Example for TLS.

Architecture support is used for misspeculation detection and recovery. First, we need hardware support to detect if a store and load actually conflict during execution. Second, if misspeculation happens, any changes to architectural state made by the speculative thread must be undone. Undoing changes to register state requires saving and restoring some register values, which can be done in software. Undoing changes to memory requires special support such as hardware transactional memory or other memory systems that buffer speculative state and discard it on misspeculation.

Next, we will give an example to show how TLS parallelizes a loop (Figure 9.18) with loop-carried dependences.

According to the PDG in Figure 9.19 for this example code, there are two loop-carried dependences. The first one is sourced by statement 4, which decides whether the loop should exit or not. The second one is in statement 6, which provides the node value for the next iteration. As for the dependence sourced by statement 3, since cost is a reduction variable, sum reduction that is mentioned in Section 9.3.3 can be used to resolve the dependence.

In order to delete all of the loop-carried dependences, we need to speculate

1. The value of cost is always less or equal to the value of T, which means the branch in statement 4 is not taken.

2. The values of nodes are the same in different invocations of the loop, which means we can use the node value in former invocations to predict the node value in the present invocation.

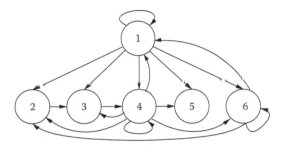

FIGURE 9.19:   PDG for example loop.

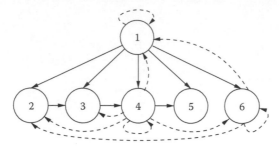

FIGURE 9.20: PDG after speculation.

```
Thread 1:

mispred  =  1;                          Thread 2:
while(node)  {
  ncost  =  calc(node);                 node  =  predicted_node;
  cost  +=  ncost;                       cost  =  0;
  if  (cost  >  T)                       while(node)  {
    break;                                 ncost  =  calc(node);
  node  =  node->next;                     cost  +=  ncost;
  if  (node  ==  predicted_node)  {       if  (cost  >  T)
    mispred  =  0;                           break;
    break;                                 node  =  node->next;
  }                                       }
}                                         send(thread1,  cost);
if  (!mispred)  {
  receive(thread2,  cost2);
  cost  +=  cost2;
}
```

FIGURE 9.21: Parallel code after TLS.

Figure 9.20 shows the PDG after speculation. The dashed lines mark the dependences having been speculated out. Now if we have two threads, we will transform the example code into the code seen in Figure 9.21.

Examining this code, the reader may have one significant question: how do we get this value for "predicted node" used by thread 2 as a starting position? One possible answer is a technique known as *speculative parallel iteration chunk execution* (Spice) [33]. Spice determines which loop-carried live-ins require value prediction. It then inserts code to gather values to be predicted. On the first invocation of the loop, it collects the values—in our case, this is a pointer to a location somewhere in the middle of our list. Since many loops in practice are invoked many times, the next invocation can then speculate that the same value will be used. In this case, so long as the predicted node was not removed from the list (even if nodes around it were removed), our speculation will succeed.

Since we have two threads here, only one node value needs to be predicted. Thread 1 is executed nonspeculatively. It keeps executing until the node value is equal to the predicted node value for thread 2. If that happens, it means there is no misspeculation, and the result from thread 2 can be combined with the result in thread 1. However, if the node value in thread 1 is never equal to the predicted value, thread 1 will execute all of the iterations and a misspeculation will be detected in the end. When misspeculation happens, all the writes to registers and memory made by thread 2 must be undone and the result of thread 2 is discarded.

There has been a great deal of work done with TLS techniques beyond what has been demonstrated in this section, beyond what could possibly be covered in this chapter. Some efforts have investigated hardware support and design for TLS techniques as with Stanford's Hydra Core Multiprocessor (CMP) [34], among many others [35–39]. Other TLS works have investigated software support for these techniques [14,32]. Bhowmik and Franklin investigate a general compiler framework for TLS techniques [42]. Other work by Kim and Yeung examines compiler algorithms [43]. Work by Johnson et al. examines program decomposition [44]. Further work by Zilles and Sohi examines a TLS technique involving master/slave speculation [45]. Readers who are interested are directed to these publications for more information about the numerous TLS techniques and support systems.

9.5 Pipeline Parallelization

9.5.1 Introduction

In previous sections, we have shown how DOALL and DOACROSS parallelization techniques are able to extract parallelism from sequential loops. In this section, we will introduce a new method called pipeline parallelization and demonstrate how speculation enables more pipeline parallelization.

A pipeline is a chain of processing elements, such as instructions, arranged in a way that each earlier element produces the input that will be used by a later element. Sometimes buffering is needed to communicate between two elements. The first pipeline parallelization technique proposed was DOPIPE. It was initially proposed alongside DOACROSS to parallelize scientific code with recurrences [46]. DOPIPE splits the original loop into several stages and spreads them among multiple threads. The number of threads are fixed at runtime. The dependences among all threads are forced to be unidirectional, which means no cyclic cross-thread dependences.

Decoupled software pipelining (DSWP) [47] is a technique proposed recently that partitions the code into several stages and executes them in a pipelined fashion. Like DOPIPE, communication is restricted to be unidirectional—from an earlier stage to a later stage in the pipeline only. DSWP differs from DOPIPE in that (1) DSWP targets general-purpose programs instead of scientific code and (2) DSWP uses communication queues for inter-thread communication. In this book, we will focus on DSWP as a state-of-the-art general-purpose pipeline parallelization technique. Figure 9.22b shows the execution model of DSWP.

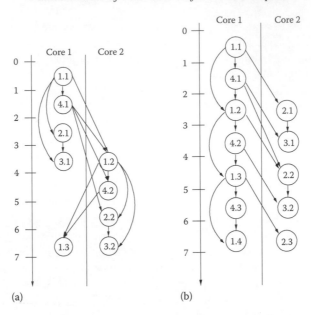

FIGURE 9.22: DOACROSS versus DSWP with communication latency. (a) DOACROSS with latency = 2 cycles. (b) DSWP with latency = 2 cycles.

Pipeline parallelization relies on the loop containing pipelineable stages. That is, DSWP works only if the loop can be split into several stages that do not form cyclic dependences in the PDG. The example code shown in Section 9.4.2 will not be a good candidate for DSWP since the two instructions contained in the loop form a cyclic data dependence. Therefore DSWP does have the same universal applicability as DOACROSS. However, DSWP is not as sensitive to communication cost as DOACROSS because the execution of the next iteration overlaps with communication. Communication latency only affects performance as a one-time cost. Therefore, pipeline parallelization is more latency tolerant than the DOACROSS technique. In Figure 9.15, we saw that DOACROSS performs poorly when the communication latency is high. DSWP allows overlapping of communication with execution of the same stage in the next iteration to hide the latency, as shown in Figure 9.22.

9.5.2 Code Partitioning

Let us again revisit the linked list traversal loop example in Figure 9.1 from the analysis section (Section 9.2).

The PDG of this loop in Figure 9.1 is seen in Figure 9.6. For purposes of further discussion, we will omit self-dependence edges—each statement in a partition will be entirely within a single partition, so these edges will inevitably be respected.

To split the loop, DSWP creates an acyclic thread dependence graph based on PDG. It first identifies strongly connected components (SCCs) in the PDG. SCC is a directed subgraph of PDG, in which there is a path from every vertex to every other vertex in

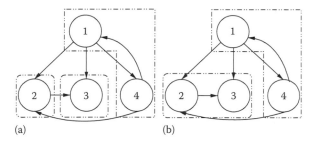

(a)   (b)

FIGURE 9.23:   DSWP code partitioning. (a) DAG_{SCC} of sample loop. (b) DAG_{SCC} partitioned.

the graph. If each of the SCCs is contracted to a single vertex, the result graph is transformed to a graph that consists of only SCCs, called DAG_{SCC} as in Figure 9.23a. After that, DAG_{SCC} is partitioned into a fixed number of threads while making sure that there is no cyclic inter-thread dependences (see Figure 9.23b). In this example, instructions 1 and 4 form an SCC that cannot be split. They have to be put on the same stage in pipeline; otherwise we will have cyclic inter-thread dependence. It is usually common to have more than one partition potentially available. Heuristics are used to guide the choice of the most appropriate partition to generate a pipeline that balances workloads and minimizes communication cost [48].

9.5.3   Code Generation

Now that we have the code partition, the next step is to generate the parallel code for the parallel threads to execute. Our example DSWP parallelization of this loop splits the loop into two slices: one linked list traversal thread and one computation thread, as shown in Figure 9.24a and b, respectively. The produce function enqueues the pointer into a communication queue, and the consume function dequeues one element (a pointer) from the queue. In this way, we communicate dependences forward in the pipeline; the isolation of thread-local variables in conjunction with this forward communication of intra-iteration dependences allows us to ignore intra-iteration anti-dependences when partitioning. If the queue is full when produce function is called, the produce will be blocked and will wait for an empty slot; if the queue is empty when consume function is called, that consuming thread will be blocked to wait on more data.

```
while (node)   {                    while (node=consume(queue))   {
    produce  (node,  queue);            ncost  =  calc(node);
    node  =  node->next;                cost  +=  ncost;
}                                   }
(a)                                 (b)
```

FIGURE 9.24:   DSWP partitioned code. (a) Produce thread. (b) Consume thread.

This DSWP parallelization decouples the execution of the two code slices and allows the code execution to overlap with the inter-thread communication. Furthermore, the loop is split into smaller slices so that the program can make better use of cache locality.

9.5.4   Advanced Topic: Speculation

In the case of DOACROSS and DOALL parallelization, we saw that applicability was limited considerably by inter-iteration dependences and may alias relations that could not be disproved at compile time. Through speculation as seen in Section 9.4.3, the applicability of DOACROSS parallelization is greatly improved in instances where cost of misspeculation detection and misspeculation recovery time can be kept relatively low in cases where frequency of misspeculation remains low.

The case for speculation in pipeline parallelization is similar. One typical case is that of a conditional branch for an error condition. Such a branch may be rarely taken, and may in fact never manifest in normal execution, but it can introduce a dependence recurrence. Consider the loop in Figure 9.25, a loop that is exactly the same as the loop examined in previous sections on pipeline parallelism except that it introduces an early exit condition if the computed cost exceeds a threshold.

As an example throughout this section, we will refer to the code in Figure 9.25. The PDG formed by this code is shown in Figure 9.26 and is clearly not partitionable by DSWP as described previously. However, were we able to remove some of these edges like the early loop exit, we can remove the dependence recurrence created by the introduction of this condition, break some SCCs, and achieve a DSWP-style

```
1:   while(node)  {
2:     ncost  =  calc(node);
3:     cost  +=  ncost;
4:     if  (cost  >  T)
5:       break;
6:     node  =  node->next;
7:   }
```

FIGURE 9.25:   Pipeline speculation code example.

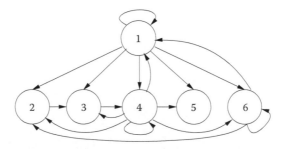

FIGURE 9.26:   PDG for Figure 9.25.

pipeline parallelization. This example demonstrates the usefulness of speculation, but leaves two important questions:

1. How do we select and break dependences to break recurrences?

2. How do we recover from incorrect speculation?

We will answer these questions in terms of the most common speculative pipeline parallelism technique, SpecDSWP [49]. The answer to the first question is in part the same answer as it was in the previous section on speculation in DOACROSS (Section 9.4.3). Possible types of speculation are the same as those seen previously. As review, these can include:

1. Value speculation—using a predicted value instead of actual value; includes biased branch speculation, load prediction, and silent store.

2. Memory alias speculation—speculating that two memory operations, one of which is a write, will not access the same memory location.

3. Memory value speculation—combining elements of both of the aforementioned and speculating that a value read from memory will be the same regardless of the memory location from which it is read.

There are, however, important differences. The most integral difference lies in the fact that TLS techniques generally speculate only on loop-carried dependences as their goal is to reduce inter-thread communication. Meanwhile, pipeline parallelization, which does not focus on reducing inter-thread latency, instead looks to remove dependence recurrences. Speculating such recurrences breaks large SCCs and reduces the size of pipeline stages, a desirable result for pipeline parallelism. This difference of focus allows speculation on *any* dependence that will break an SCC. Since loop-carried dependences may not be easily speculatable while another dependence along the cycle may be, the compiler has increased flexibility when selecting which edges to speculate.

The fact that speculation in pipeline parallelizations can be intra-iteration speculation has other implications. Unlike DOACROSS speculation techniques, no iteration executes nonspeculatively. The resultant work is committed when all stages of the pipeline have completed without misspeculation. The iterations must be guaranteed to commit in proper order.

Though it is now clear that it is desirable to select dependence edges to speculate which will remove a recurrence with the least likelihood of causing excessive misspeculation, the question of how to select these dependences remains. Limiting the likelihood of excessive misspeculation can be done using *profiling*.* Loop-sensitive profiling provides even better information to a compiler looking to limit misspeculation [50]. Representative loop-sensitive profiling information can indicate rarely

---

* Profiling involves instrumenting the original code with a code to collect data at run time. The instrumented code is then compiled and run on representative sets of data. Information collected from the instrumented code can be used as a supplement for analysis as it provides information that may not be determined practically (or at all) by static analysis.

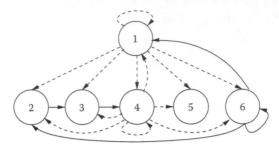

FIGURE 9.27:   Prospectively speculated dependences for Figure 9.26.

taken control edges, possible memory aliases that rarely or never manifest, and operations that frequently return the same result. If profiling information is not available, it
has been found that early loop exit branches are often good candidates for speculation
as loops generally execute many times before exiting in practice.

In the example in Figure 9.25, let us assume we know that the early exit for cost
greater than a threshold (`cost > T`) is rarely taken and thus a good candidate for
speculation. We will also assume that the exit due to a null node is also rarely taken
(profiling reveals we have long lists) and thus a possible candidate for speculation as
well. We see in Figure 9.27 the dashed lines representing the candidates for speculated
dependences.

The second issue in selecting dependences is actually choosing dependences that
break a recurrence. It may be, and in fact often is, the case that it is necessary to speculate more than one dependence in order to break a recurrence. Ideally, one would want
to consider speculating all sets of dependences. Unfortunately, exponentially many
dependence sets exist and so a heuristic solution is necessary. In SpecDSWP, this is
handled in a multistep manner. After determining which edges are easily speculatable, all of these edges are prospectively removed. The PDG is then partitioned in the
same manner as in DSWP with these edges removed. Edges that have been prospectively removed and would break an inter-thread recurrence that originates later in
the pipeline become speculated. Other edges that do not fit these criteria are effectively returned to the PDG. That is, the dependences are respected and not actually
speculated.

In the example as seen in Figure 9.28, we see that DSWP partitioning algorithm
has elected to place nodes 1 and 6 on a thread, statement 2 on a thread, and statements 3, 4, and 5 on a thread. Using this partitioning, it is clear that speculating the
dependences on the early exit due to cost exceeding a threshold (outbound edges
from 4) is necessary. However, edges speculated due to a null node do not break any
inter-thread recurrences. These dependences, therefore, will not be speculated.

Though the previous paragraphs describe what is necessary to select dependences
to speculate, the question of how to check for misspeculation must still be resolved.
This is partly a problem of code generation and partly a problem of support systems.
Unlike speculation in DOACROSS techniques (i.e., TLS), every iteration is speculative and no single thread executes a whole iteration. This means that there must
be some commit unit to guarantee a proper order of commit and handle recovery.

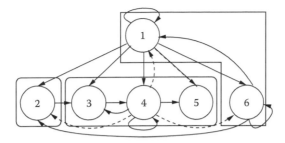

FIGURE 9.28: Partitioned PDG for Figure 9.25.

```
                                                         while (TRUE)  {
                           while (TRUE)  {                  ncost  =  consume();
while (node)  {              node  =  consume();            cost  +=  ncost;
  node  =  node->next;       ncost  =  calc(node);         if  (cost  >  T)
  produce(node);             produce(ncost);                  FLAG_MISSPECULATION();
}                          }                             }

      Thread 1                    Thread 2                    Thread 3
```

FIGURE 9.29: Parallelized code from Figure 9.25.

Furthermore, the memory system must allow multiple threads to execute inside the same iteration at one time. The system must check for proper speculation across all the threads in the same iteration before it can be committed. These checks for misspeculation may either be inserted into the code by the compiler or handled in a hardware memory system. SpecDSWP handles most speculation types by inserting checks to flag misspeculation. The noteworthy exception is the case of memory alias speculation where the memory system is left to detect cases where a memory alias did exist.

In our example, we will insert code to flag misspeculation if we have, in fact, misspeculated. We will place this misspeculation flagging code when we can determine if misspeculation has occurred; this will be in the final thread. Three threads will be generated from the partitioned PDG in the same way as DSWP. The code generated is shown in Figure 9.29.

In all cases, the final requirement is a method by which speculative state can be buffered prior to commit and a unit to commit nonspeculative state. When misspeculation is signaled, the nonspeculative state must be recovered and the work must be recomputed. Version memory systems [51] that allow multiple threads to operate on a given version and commit only when the version has been completed and well speculated have been proposed in hardware and in software but will not be discussed here.

The considerations made in this section can be summarized as a step-by-step procedure. In this case, we can break the overall procedure here into six steps:

1. Build the PDG for the loop to be parallelized.

2. Select the dependence edges to speculate.

3. Remove the selected edges from the PDG.

4. Apply the DSWP transformation to the PDG without speculated edges.

5. Insert code necessary to detect misspeculation for dependences to be speculated.

6. Insert code for recovery from misspeculation.

A more detailed explanation of these steps can be found in work by Vachharajani [51].

9.6   Bringing It All Together

In the last three sections, we have seen a variety of techniques that may be used to take advantage of parallelizable sequential code. We saw a highly scalable technique in DOALL, a universally applicable technique in DOACROSS, and a widely applicable, latency-tolerant technique in pipelined parallelism with DSWP. We have also seen the use of speculation to increase opportunities for parallelism. In this section, we bring together some of the best aspects of these techniques in the form of a technique called Speculative Parallel Stage Decoupled Software Pipelining (Spec-PS-DSWP).

Spec-PS-DSWP is based on the pipelining execution model of DSWP with the use of speculation to break recurrences as in SpecDSWP and combines them with an idea that partially mirrors DOALL by replicating stages that can be executed independently. The result is a highly applicable technique that is latency tolerant and highly scalable. We refer to the code example in Figure 9.30 to demonstrate how we can bring all of these elements together.

If pointer p is local to each iteration and analysis can determine that the inner lists traversed using p do not alias with the outer list traversed using node, the compiler can build the PDG shown in Figure 9.31a.

Examining the code, there are control dependences emanating from the early loop exit condition in the inner loop that create a dependence recurrence much as we saw

```
1:   while(node)  {
2:      p  =  node->list;  //  'p'  is  iteration  local
3:      while(p)  {
4:         if  (p->val  ==  0)  exit();
5:         p->val++;
6:         p  =  p->next;
      }
7:      node  =  node->next;
   }
```

FIGURE 9.30:   Code example for Spec-PS-DSWP.

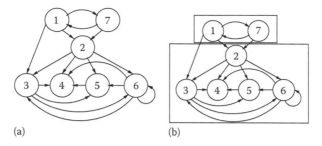

(a)   (b)

FIGURE 9.31: PDG of sample code (a) before partitioning and (b) after partitioning.

in Section 9.5.4. With this control dependence removed via speculation, the compiler can create a two-stage pipeline as before. Figure 9.31b shows the partitioned PDG.

However, the second stage is likely to be much longer than the first stage—it contains a whole linked-list traversal of its own. This imbalance greatly diminishes the gains from a SpecDSWP parallelization. The fix for this is related to the concept of independence we saw before with DOALL in Section 9.3. Note that the invocations of the inner loop (lines 3–6) are independent of one another, if memory analysis proves that each invocation of the inner loop accesses different linked lists. This stage could be replicated and all invocations of this loop could be executed in parallel. This *parallel stage* can be executed in parallel on many cores and can be fed with the nodes found in the first *sequential stage* of the pipeline, as shown in Figure 9.32. The parallel stages threads need not communicate with each other, just as DOALL threads need not communicate with each other. Their only common link is the first DSWP pipeline stage. Because the parallel stage can be replicated many times, Spec-PS-DSWP is a highly scalable technique, achieving better speedup as the number of cores increases.

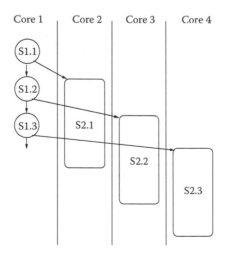

FIGURE 9.32: Execution model of Spec-PS-DSWP.

9.7 Conclusion

Single-threaded programs cannot make use of the additional cores on a multicore machine. Thread-level parallelism (TLP) must be extracted from programs to achieve meaningful performance gains. Automatic parallelization techniques that extract threads from single-threaded programs expressed in the sequential programming model relieve the programmer of the burden of identifying the parallelism in the application and orchestrating the parallel execution on different generations of target architectures. This solution appears even more appealing in the face of immense programmer productivity loss due to the limited tools to write, debug, and performance-optimize multi-threaded programs. Automatic parallelization promises to allow the programmer to focus attention on developing innovative applications and providing a richer end-user experience, rather than on optimizing the performance of a piece of software.

Automatic parallelization techniques focus on extracting loop-level parallelism. Executing each iteration of a loop concurrently with others (DOALL) is a highly scalable strategy, but few general-purpose loops have dependence patterns amenable to such execution. In the face of loop-carried dependences, alternative schemes such as DOACROSS and DSWP have been developed. Like DOALL, DOACROSS schedules independent iterations for concurrent execution while synchronizing dependent iterations. However, DOACROSS puts the inter-thread communication latency on the critical path as seen in Figure 9.15. DSWP addresses this problem by partitioning the loop body in such a way as to keep dependence recurrences local to a thread and communicate only off-critical path dependences in a unidirectional pipeline as in Figure 9.22. However, DSWP's scalability is restricted by the balance among, and the number of, pipeline stages. PSDSWP [52] addresses this issue: It combines the applicability of DSWP with the scalability of DOALL by replicating a large pipeline stage across multiple threads, provided the stage has no loop-carried dependences with respect to the loop being parallelized. All of these techniques depend on compile-time dependence analyses that, unfortunately, become very complicated on programs that exhibit complex memory access patterns. Worse yet, some dependences can be discovered only at runtime. Speculative execution has been proposed to overcome these limitations. Thread-level speculation (TLS) techniques allow the compiler to optimistically ignore statically ambiguous dependences, by relying on a runtime system to detect misspeculation and resume execution from a nonspeculative program state.

Historically, automatic parallelization was successful primarily in the numeric and scientific domains on well-behaved, regular programs written in Fortran-like languages [53,54]. The techniques discussed in this chapter expand the scope of automatic parallelization by successfully extracting parallelism from general-purpose programs, such as those from the SPEC benchmark suite, written in Fortran, C, C++, or Java [4,37,55–58].

In order to speed up a wider range of applications, a modern auto-parallelizing compiler must possess an arsenal of parallelization techniques and optimizations. DSWP and its extensions, SpecDSWP and PSDSWP, have proven to be robust and scalable and will form an integral part of the arsenal. Traditional data-flow analyses to compute register and control dependences and state-of-the-art memory dependence analyses will be needed to statically disprove as many dependences as possible. Additionally, to be able to target loops at any level in the program, from innermost to outermost, it is necessary that all analyses and techniques have interprocedural scope. Speculation, synchronization placement, and other optimizations require accurate loop-aware profile information. Branch profilers, silent store value profilers, memory alias profilers, etc., will all be part of the compilation framework. To support speculation, runtime support either in software or hardware will be necessary. This layer will be responsible for detecting misspeculation (the checks may be inserted at compile time) and rolling back program state. In summary, such a tool-chain will allow a software developer to continue developing in a sequential programming model using the robust development environments of today while still obtaining scalable performance on the multicore and manycore architectures of tomorrow.

Future parallelization systems must be capable of dynamically adapting the compiled application to the underlying hardware and the execution environment. Instead of complicating the static optimization phase and incorporating all possible dynamic scenarios, a better solution is to assume a generic runtime configuration and perform aggressive optimizations statically while simultaneously embedding useful metadata (hints) into the binary. A runtime system can then use this metadata to adapt the code to the underlying operating environment. Adaptation to resource constraints, execution phases in the application, and dynamic dependence patterns will be necessary. As the number of cores on a processor grows, one or more cores may be dedicated to the tasks of collecting profile information and performing dynamic program transformation. To reduce the overheads of speculation and profiling, significant advances in memory dependence analysis must be made to inform the compiler that some dependences do not exist and hence need not be profiled and speculated. Research in memory shape analysis [14–16] promises to be an interesting avenue to address this issue.

Automatic thread extraction will not be restricted just to sequential programs; as many parallel programs have been and will continue to be written, individual threads of these programs must be further parallelized if they are to truly scale to hundreds of cores. Furthermore, just as modern optimizing ILP compilers include a suite of optimizations with many of them interacting in a synergistic fashion, so must the various parallelization techniques be applied synergistically, perhaps hierarchically, at multiple levels of loop nesting, to extract maximal parallelism across a wide spectrum of applications.

References

1. J.C. Corbett. Evaluating deadlock detection methods for concurrent software. *IEEE Transactions on Software Engineering*, 22(3):161–180, 1996.

2. C. Demartini, R. Iosif, and R. Sisto. A deadlock detection tool for concurrent Java programs. *Software: Practice and Experience*, 29(7):577–603, 1999.

3. G.R. Luecke, Y. Zou, J. Coyle, J. Hoekstra, and M. Kraeva. Deadlock detection in MPI programs. *Concurrency and Computation: Practice and Experience*, 14(11):911–932, 2002.

4. M.J. Bridges, N. Vachharajani, Y. Zhang, T. Jablin, and D.I. August. Revisiting the sequential programming model for the multicore era. *IEEE Micro*, 28(1), January 2008.

5. S. Horwitz, T. Reps, and D. Binkley. Interprocedural slicing using dependence graphs. *ACM Transactions on Programming Languages and Systems*, 12(1):26–60, 1990.

6. S.S. Muchnick. *Advanced Compiler Design and Implementation*. Morgan Kaufmann Publishers Inc., San Francisco, CA, 1997.

7. A. Orso, S. Sinha, and M.J. Harrold. Classifying data dependences in the presence of pointers for program comprehension, testing, and debugging. *ACM Transactions on Software Engineering and Methodology*, 13(2):199–239, 2004.

8. D.E. Maydan, J.L. Hennessy, and M.S. Lam. Effectiveness of data dependence analysis. *International Journal of Parallel Programming*, 23(1):63–81, 1995.

9. K. Psarris and K. Kyriakopoulos. The impact of data dependence analysis on compilation and program parallelization. In *ICS '03: Proceedings of the 17th Annual International Conference on Supercomputing*, pp. 205–214, New York, 2003. ACM, New York.

10. A.V. Aho, R. Sethi, and J.D. Ullman. *Compilers: Principles, Techniques, and Tools*. Addison-Wesley Longman Publishing Co., Inc., Boston, MA, 1986.

11. M. Hind. Pointer analysis: Haven't we solved this problem yet? In *PASTE '01: Proceedings of the 2001 ACM SIGPLAN-SIGSOFT Workshop on Program Analysis for Software Tools and Engineering*, pp. 54–61, New York, 2001. ACM, New York.

12. D.J. Pearce, P.H.J. Kelly, and C. Hankin. Efficient field-sensitive pointer analysis of C. *ACM Transactions on Program Languages and Systems*, 30(1):4, 2007.

13. B.-C. Cheng and W.-M.W. Hwu. Modular interprocedural pointer analysis using access paths: Design, implementation, and evaluation. In *PLDI '00: Proceedings of the 2000 ACM SIGPLAN Conference on Programming Language Design and Implementation*, pp. 57–69, New York, 2000. ACM, New York.

14. M. Sagiv, T. Reps, and R. Wilhelm. Parametric shape analysis via 3-valued logic. In *POPL '99: Proceedings of the 26th ACM SIGPLAN-SIGACT Symposium on Principles of Programming languages*, pp. 105–118, New York, 1999. ACM, New York.

15. B. Guo, N. Vachharajani, and D.I. August. Shape analysis with inductive recursion synthesis. In *PLDI '07: Proceedings of the 2007 ACM SIGPLAN Conference on Programming Language Design and Implementation*, pp. 256–265, New York, 2007. ACM, New York.

16. C. Calcagno, D. Distefano, P. O'Hearn, and H. Yang. Compositional shape analysis by means of bi-abduction. In *POPL '09: Proceedings of the 36th Annual ACM SIGPLAN-SIGACT Symposium on Principles of Programming Languages*, pp. 289–300, New York, 2009. ACM, New York.

17. K. Kennedy and J.R. Allen. *Optimizing Compilers for Modern Architectures: A Dependence-Based Approach*. Morgan Kaufmann Publishers Inc., San Francisco, CA, 2002.

18. W. Pugh. A practical algorithm for exact array dependence analysis. *Communications of the ACM*, 35(8):102–114, 1992.

19. D.A. Padua and Y. Lin. Demand-driven interprocedural array property analysis. In *LCPC '99*, pp. 303–317, London, U.K., 1999.

20. U. Banerjee. Data dependence in ordinary programs. Master's thesis, University of Illinois at Urbana-Champaign, Champaign, IL, 1976.

21. V. Sarkar and S.J. Fink. Efficient dependence analysis for Java arrays. In *Euro-Par '01: Proceedings of the 7th International Euro-Par Conference Manchester on Parallel Processing*, pp. 273–277, London, U.K., 2001. Springer-Verlag, Berlin, Germany.

22. J. Ferrante, K.J. Ottenstein, and J.D. Warren. The program dependence graph and its use in optimization. *ACM Transactions on Programing Languages and System*, 9(3):319–349, 1987.

23. A.J. Bernstein. Analysis of programs for parallel processing. *IEEE Transactions on Electronic Computers*, 15(5):757–763, October 1966.

24. R. Allen and K. Kennedy. *Optimizing Compilers for Modern Architectures*. Morgan Kaufmann, San Francisco, CA, 2002.

25. H. Zhong, M. Mehrara, S. Lieberman, and S. Mahlke. Uncovering hidden loop level parallelism in sequential applications. In *HPCA'08: Proceedings of the 14th International Symposium on High-Performance Computer Architecture (HPCA)*, Ann Arbor, MI, February 2008.

26. J. Allen and K. Kennedy. Automatic loop interchange. In *Proceedings of the ACM SIGPLAN '84 Symposium on Compiler Construction*, pp. 233–246, Montreal, Quebec, Canada, June 1984.

27. Q. Yi and K. Kennedy. Improving memory hierarchy performance through combined loop interchange and multi-level fusion. *International Journal of High Performance Computing Applications*, 18(2):237–253, 2004.

28. A. Wakatani and M. Wolfe. A new approach to array redistribution: Strip mining redistribution. In *Proceedings of Parallel Architectures and Languages Europe (PARLE 94)*, pp. 323–335, London, U.K., 1994.

29. A.W. Lim and M.S. Lam. Maximizing parallelism and minimizing synchronization with affine partitions. *Parallel Computing*, 24(3–4):445–475, 1998.

30. S.P. Midkiff and D.A. Padua. Compiler algorithms for synchronization. *IEEE Transactions on Computers*, 36(12):1485–1495, 1987.

31. P. Tang, P.-C. Yew, and C.-Q. Zhu. Compiler techniques for data synchronization in nested parallel loops. *SIGARCH Computer Architecture News*, 18(3b):177–186, 1990.

32. H.-M. Su and P.-C. Yew. Efficient doacross execution on distributed shared-memory multiprocessors. In *Supercomputing '91: Proceedings of the 1991 ACM/IEEE Conference on Supercomputing*, pp. 842–853, New York, 1991. ACM, New York.

33. E. Raman, N. Vachharajani, R. Rangan, and D.I. August. Spice: Speculative parallel iteration chunk execution. In *Proceedings of the 2008 International Symposium on Code Generation and Optimization*, New York, 2008.

34. L. Hammond, B.A. Hubbert, M. Siu, M.K. Prabhu, M. Chen, and K. Olukotun. The Stanford Hydra CMP. *IEEE Micro*, 20(2):71–84, January 2000.

35. H. Akkary and M.A. Driscoll. A dynamic multithreading processor. In *Proceedings of the 31st Annual ACM/IEEE International Symposium on Microarchitecture*, pp. 226–236, Los Alamitos, CA, 1998. IEEE Computer Society Press, Washington, DC.

36. P. Marcuello and A. González. Clustered speculative multi-threaded processors. In *Proceedings of the 13th International Conference on Supercomputing*, pp. 365–372, New York, 1999. ACM Press, New York.

37. J.G. Steffan, C. Colohan, A. Zhai, and T.C. Mowry. The STAMPede approach to thread-level speculation. *ACM Transactions on Computer Systems*, 23(3):253–300, February 2005.

38. J.-Y. Tsai, J. Huang, C. Amlo, D.J. Lilja, and P.-C. Yew. The superthreaded processor architecture. *IEEE Transactions on Computers*, 48(9):881–902, 1999.

39. T.N. Vijaykumar, S. Gopal, J.E. Smith, and G. Sohi. Speculative versioning cache. *IEEE Transactions on Parallel and Distributed Systems*, 12(12):1305–1317, 2001.

40. M. Cintra and D.R. Llanos. Toward efficient and robust software speculative parallelization on multiprocessors. In *PPoPP '03: Proceedings of the Ninth ACM SIGPLAN Symposium on Principles and Practice of Parallel Programming*, pp. 13–24, New York, 2003. ACM, New York.

41. C.E. Oancea and A. Mycroft. Software thread-level speculation: An optimistic library implementation. In *IWMSE '08: Proceedings of the 1st International Workshop on Multicore Software Engineering*, pp. 23–32, New York, 2008. ACM, New York.

42. A. Bhowmik and M. Franklin. A general compiler framework for speculative multithreading. In *Proceedings of the 14th ACM Symposium on Parallel Algorithms and Architectures*, pp. 99–108, New York, August 2002.

43. D. Kim and D. Yeung. A study of source-level compiler algorithms for automatic construction of pre-execution code. *ACM Transactions on Computing Systems*, 22(3):326–379, 2004.

44. T.A. Johnson, R. Eigenmann, and T.N. Vijaykumar. Min-cut program decomposition for thread-level speculation. In *Proceedings of the ACM SIGPLAN 2004 Conference on Programming Language Design and Implementation*, pp. 59–70, Washington, DC, June 2004.

45. C. Zilles and G. Sohi. Execution-based prediction using speculative slices. In *Proceedings of the 28th International Symposium on Computer Architecture*, New York, NY, July 2001.

46. D.A. Padua Haiek. Multiprocessors: Discussion of some theoretical and practical problems. PhD thesis, Champaign, IL, 1980.

47. G. Ottoni, R. Rangan, A. Stoler, and D.I. August. Automatic thread extraction with decoupled software pipelining. In *Proceedings of the 38th IEEE/ACM International Symposium on Microarchitecture*, pp. 105–116, Barcelona, Spain, November 2005.

48. G. Ottoni and D.I. August. Communication optimizations for global multithreaded instruction scheduling. In *Proceedings of the 13th ACM International Conference on Architectural Support for Programming Languages and Operating Systems*, pp. 222–232, Seattle, WA, March 2008.

49. N. Vachharajani, R. Rangan, E. Raman, M.J. Bridges, G. Ottoni, and D.I. August. Speculative decoupled software pipelining. In *Proceedings of the 16th International Conference on Parallel Architecture and Compilation Techniques*, Washington, DC, September 2007.

50. E. Raman. Parallelization techniques with improved dependence handling. PhD thesis, Department of Computer Science, Princeton University, Princeton, NJ, June 2009.

51. N. Vachharajani. Intelligent speculation for pipelined multithreading. PhD thesis, Department of Computer Science, Princeton University, Princeton, NJ, November 2008.

52. E. Raman, G. Ottoni, A. Raman, M. Bridges, and D.I. August. Parallel-stage decoupled software pipelining. In *Proceedings of the 2008 International*

Symposium on Code Generation and Optimization, Princeton University, Princeton, NJ, April 2008.

53. B. Blume, R. Eigenmann, K. Faigin, J. Grout, J. Hoeflinger, D. Padua, P. Pettersen, B. Pottenger, L. Rauchwerger, P. Tu, and S. Weatherford. Polaris: The next generation in parallelizing compilers. In *Proceedings of the Workshop on Languages and Compilers for Parallel Computing*, pp. 10.1–10.18, Ithaca, NY, August 1994. Springer-Verlag, Berlin/Heidelberg.

54. M.W. Hall, S.P. Amarasinghe, B.R. Murphy, S.-W. Liao, and M.S. Lam. Interprocedural parallelization analysis in suif. *ACM Transactions on Programming Languages and Systems*, 27(4):662–731, 2005.

55. M.K. Chen and K. Olukotun. The JRPM system for dynamically parallelizing java programs. *SIGARCH Computer Architecture News*, 31(2):434–446, 2003.

56. M.K. Prabhu and K. Olukotun. Exposing speculative thread parallelism in SPEC2000. In *Proceedings of the Tenth ACM SIGPLAN Symposium on Principles and Practice of Parallel Programming*, pp. 142–152, New York, 2005. ACM Press, New York.

57. A. Raman, H. Kim, T.R. Mason, T.B. Jablin, and D.I. August. Speculative parallelization using software multithreaded transactions. In *Proceedings of the Fifteenth International Conference on Architectural Support for Programming Languages and Operating Systems (ASPLOS)*, New York, March 2010.

58. L. Rauchwerger and D. Padua. The LRPD test: Speculative run-time parallelization of loops with privatization and reduction parallelization. *ACM SIGPLAN Notices*, 30(6):218–232, 1995.

59. SPEC Benchmarks. The standard performance evaluation corporation. www.spec.org, 1997.

Chapter 10

Auto-Tuning Parallel Application Performance

Christoph A. Schaefer, Victor Pankratius, and Walter F. Tichy

Contents

10.1   Introduction

Software needs to be parallelized to exploit the performance potential of multi-core chips. Unfortunately, parallelization is difficult and often leads to disappointing results. A myriad of parameters such as choice of algorithms, number of threads, size of data partitions, number of pipeline stages, load balancing technique, and other issues influence parallel application performance. It is difficult for programmers to set these parameters correctly. Moreover, satisfactory choices vary from platform to platform: Programs optimized for a particular platform may have to be retuned for others. Given the number of parameters and the growing diversity of multicore architectures, a manual search for satisfactory parameter configurations is impractical. This chapter presents a set of techniques that find near-optimal parameter settings automatically, by repeatedly executing applications with different parameter settings and searching for the best choices. This process is called auto-tuning.

10.2   Motivating Example

Figure 10.1 sketches the architecture of a commercial chemical data analysis application. The architecture consists of three nested parallel layers. The first layer (pipeline layer) is a software pipeline. Each one of the four stages runs in parallel. Stages 2 and 3 of the pipeline themselves contain several parallel modules (module layer). Modules M_1 to M_{10} implement different algorithms that execute in an order determined by data dependencies. On the third level, each module itself employs data parallelism (data layer).

Not only is this architecture complicated, it also requires fine-tuning. At the upper two levels, an important tuning issue is whether to use individual threads for every pipeline stage, or whether to fuse some of the stages in order to avoid buffering and synchronization overheads. Conversely, if certain stages turn out to be bottlenecks, it might pay to replicate these stages. At the lowest level, relevant tuning issues are how far to partition the data, how many threads to use, and how to balance the work. The best choice for all of these parameters depends on the relative speeds and scalability of the various modules, number of available processor cores, presence and efficiency of hyper-threading, cache performance, memory bandwidth, I/O performance, and other considerations. Obviously, making these choices involves trial and error.

Automatic performance tuning (auto-tuning) can help here. Auto-tuning not only finds near-optimal configurations, but also makes applications performance-portable. Moreover, maintenance work is reduced, because re-tuning is automated.

Is auto-tuning worth the effort? For the chemical analysis application discussed earlier, the difference in runtime between best and worst configuration is a factor of 2.4, for an overall speedup of 3.1 on eight processors. Examples later in this chapter show improvements of factors of four to five by auto-tuning. Auto-tuning is a major contributor to speedups. Given the many performance-critical tuning parameters, it may become indispensable for complex parallel programs [22,25].

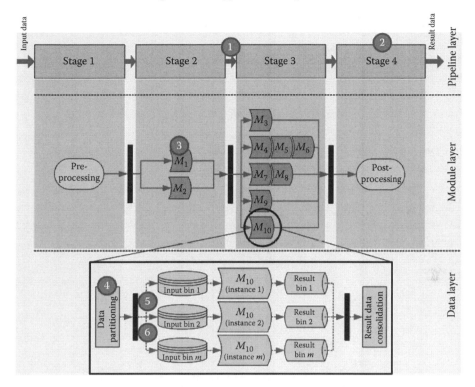

FIGURE 10.1: Architecture outline of a parallel chemical data analysis application. The most important tuning parameters are: (1) Data-parallel/sequential stages, (2) number of stages, (3) choice of algorithms, (4) size of data partitions, (5) load balancing strategy, and (6) number of algorithm instances.

In the past, most auto-tuning work has focused on the properties of numerical applications [7,29]. The methods applied are highly specialized and require significant programming effort to prepare applications for tuning. We present a broader approach that tunes general-purpose applications. This approach is highly automated, requiring merely a short specification of the parallel architecture of the application to be tuned. All tuning-related code is generated automatically or supplied by libraries. Of course, the tuning process itself is also fully automated. The rest of the chapter discusses this approach and how it compares to other work.

10.3 Terminology

This section provides an overview of auto-tuning terminology and a classification of various auto-tuning approaches.

10.3.1 Auto-Tuning

Tuning parameter: A tuning parameter p represents a program variable with a value set V_p, where $v \in V_p$ is called a parameter value of p.

Adjusting p by assigning different parameter values, $v \in V_p$ may have an impact on program performance.

Parameter configuration: A parameter configuration C is a tuple consisting of a value for each tuning parameter within a given program. Let $P = \{p_1, p_2, \ldots, p_n\}$ be the finite set of all tuning parameters in the program. Then a configuration of C is an n-tuple (v_i), where $v_i \in V_{p_i}$ and $i \in \{1, \ldots, n\}$.

Search space: The search space $\mathcal{S}$ of an auto-tuner is the cartesian product of all tuning parameter value sets of a given program. Let $P = \{p_1, p_2, \ldots, p_n\}$ be the set of all tuning parameters of a program and V_{p_i} the value set of p_i (with $i \in \{1, 2, \ldots, n\}$). Then the search space is defined as:

$$\mathcal{S} = V_{p_1} \times V_{p_2} \times \ldots \times V_{p_n} \tag{10.1}$$

The size $|\mathcal{S}|$ of the search space (i.e., the number of parameter configurations in $\mathcal{S}$) is given as

$$|\mathcal{S}| = |V_{p_1}| \cdot |V_{p_2}| \cdot \ldots \cdot |V_{p_n}| \tag{10.2}$$

Performance metric: A performance metric provides a value for the performance of a given configuration C. A common performance metric is execution time; other options are memory consumption, energy use, or response time. The auto-tuner searches for a configuration that optimizes performance under a given metric.

Measuring point: A measuring point denotes a particular point within the program, where data is collected to measure performance. A measuring point depends on a particular performance metric. A program may contain any number of measuring points. Our approach requires a rule for combining multiple performance values (e.g., by computing averages).

Performance value: The performance value $\ell \in \mathbb{R}$ of a program $\wp$ is defined as

$$\ell_\wp = \omega_\wp(p_1, \ldots, p_n) : K \to \mathbb{R}, K \in \mathcal{S} \tag{10.3}$$

For a given program $\wp$, the function $\omega_\wp$ maps a parameter configuration C on a scalar value, the performance value $\ell_\wp$. Thus, each parameter configuration $K \in \mathcal{S}$ is assigned a performance value.

Optimization: We employ an empirical approach to find the minimum or maximum performance value $\ell_\wp$ of a program $\wp$. That is, we search for the extremum of the function $\omega_\wp$.

For example, if we want to optimize the program's execution time, we search for the parameter configuration that results in the shortest execution time.

In many practical scenarios, the function $\omega_\wp$ and its gradient $\nabla \omega_\wp$ are unknown. This is the main reason why auto-tuning techniques are required at all. In case $\omega_\wp$

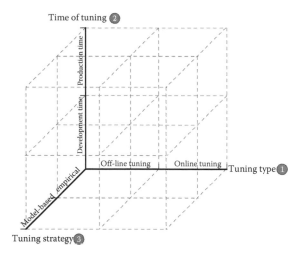

FIGURE 10.2: Taxonomy of auto-tuning approaches.

would be differentiable and $\nabla \omega_{\wp}$ known, we could employ a number of algorithms to solve the problem analytically. If that were possible, we would know the impact of the tuning parameters and their dependencies and $\omega_{\wp}$ would represent an analytical model for the program's performance. However, it is far from easy to create analytical models that are able to precisely predict the performance of entire applications on different target platforms. That is why we rely on a empirical, search-based optimization approach [31].

10.3.2 Classification of Approaches

The field of automatic performance tuning covers a wide range of different approaches. Figure 10.2 shows an *auto-tuning taxonomy*. Each dimension of the taxonomy cube characterizes one of the following specifies:

Tuning type (dimension 1): The *tuning type* defines when the values of the tuning parameters are modified. We distinguish between *off-line tuning* and *online tuning*.

- *Off-line tuning* modifies parameter values between separate program runs. A single step in an off-line auto-tuner determines a parameter configuration, inserts it into the program, and then starts and monitors the configured program to get a performance value.

- *Online tuning* modifies parameter values while the program is running. Within a single program run, several parameter configurations can be sampled. The online approach requires some preconditions. First, the application must be programmed in a way that parameter modifications at run-time are possible and do not lead to inconsistent states. Second, there must be a structure in which a part of the program (e.g., a loop performing a significant amount of work) is repeated over and over again. The program must also run long enough to allow an auto-tuner to sample enough parameter configurations in a single execution. If there is not enough time, online and off-line tuning can be combined.

Time of tuning (dimension 2): The *time of tuning* specifies when the auto-tuner is active. We distinguish between development time and production time.

- *Development time*: Tuning runs are performed during the final test phase of the development process. The auto-tuner becomes a development tool finding the best parameter configuration for a specific target platform.

- *Production time*: Tuning runs are performed during initial production runs, or periodically for re-tuning. Tuning may slow down the application because of data gathering and because unfavorable configurations must be tried out.

Development and production time tuning can be combined. Tuning at development time can provide a configuration that is a good starting point on a certain class of platforms and certain data sets, while tuning during production time fine-tunes the configuration to the local situation. Both off-line and online tuning can be used.

Search strategy (dimension 3): The *search strategy* finds the best parameter configuration.

- *Empirical tuning*: Empirical (or search-based) tuning executes the program repeatedly on typical inputs to sample performance under different parameter configurations. It employs a meta-heuristics (see Section 10.7.2) to traverse the search space. Empirical tuning is applicable to a wide range of applications. However, managing the exponentially growing search space is challenging. Concepts for search space reduction are indispensable and will be discussed in Section 10.6.

- *Model-based tuning*: Model-based tuning employs analytic performance models to predict program performance under various configurations. In contrast to the empirical approach, model-based tuning does not search for the best parameter configuration by sampling, but predicts the optimal parameter values based on the model. Models are typically adapted to a specific application domain, program structure, hardware platform, or programming model. If one of these characteristics changes, the model needs to be re-adapted. Some of the adaption can be done by sampling (off-line or online).

10.4   Overview of the Tunable Architectures Approach

Figure 10.3 illustrates the steps for creating a parallel application with tunable architectures. We use exclusively offline, empirical auto-tuning during development time. The process starts with designing the parallel architecture using certain patterns

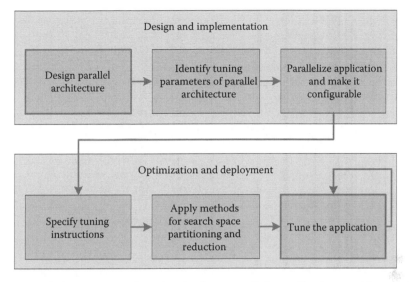

FIGURE 10.3: Process for developing parallel applications with tunable architectures.

such as master/worker, pipeline, and data decomposition. Next, performance-relevant parameters are identified. Then the application is actually parallelized and the identified tuning parameters are made adjustable.

The actual implementation of the tunable parallel program is completed with these three steps. However, the application is not yet *automatically* tunable. We need to provide a separate component, an empirical auto-tuner, and inform it about the tuning parameters and their ranges. This information is specified using an instrumentation language, which is presented later.

Each of the steps is complex, so we will illustrate the respective tasks in more detail in the following sections.

10.5 Designing Tunable Applications

Auto-tuning must take the architecture of an application into account. As we have seen in Section 10.2, a large application may contain several parallel sections in multiple layers. The chemical data analysis application combines several parallel patterns, such as pipeline parallelism, task parallelism, and data parallelism. Its architecture offers many tuning parameters. Which pipeline stages should have their own thread, which ones should be fused? How many instances of a stateless algorithm should be created? Which load-balancing strategy would be the best? How does the target platform influence the choices?

Earlier studies [21,22,24] have shown that coarse-grained software architecture adaptations can provide significant speedup gains. However, adaptations of the program's architecture are rarely done because they require invasive and complex code changes. Developers are hesitant to make such changes because they prefer to avoid the risk of introducing bugs.

Automated software architecture adaptation can be based on using *parallel patterns* as building blocks. Similar to design patterns introduced in [8], a parallel pattern describes a solution for a class of parallelization problems. Examples of commonly used patterns are the pipeline pattern and the fork-join pattern. A pattern's behavior and its specific performance bottlenecks can be studied and this knowledge can be exploited to (1) equip patterns with tuning parameters and (2) use patterns as predefined configurable building blocks for a parallel program. We call parallel patterns exposing tuning parameters *tunable patterns*.

We employ a predefined set of tunable patterns to define a program's architecture. Compared to low-level parallelization techniques (not normally considered during the design phase of a program), this approach offers several advantages: a structured top-down design, modularity, and the exploitation of coarse-grained parallelism.

The following section describes our tunable architectures approach and CO^2P^3S, a related approach from the literature.

10.5.1   Tunable Architectures

Our tunable architectures approach [26] allows designing parallel applications that are parallel and auto-tunable. The approach concentrates on shared-memory multicore platforms.

The concept of tunable architectures is based on the *tunable architecture description language* (TADL) that provides operators to express parallel patterns. At the outset, a developer defines atomic software components. Atomic components are opaque to the auto-tuner: they contain nothing an auto-tuner needs to adjust. Typically, though not necessarily, atomic components execute sequentially. Next, the developer uses pattern operators to combine atomic components into parallel patterns. All pattern operators implicitly expose a predefined set of performance parameters which will be adjusted by an auto-tuner. TADL also supports the declaration of alternative architectures that an auto-tuner can try out.

10.5.1.1   Atomic Components

An atomic component represents an elementary work item of a program. It does not provide auto-tunable parallelism, but is intended to run in parallel with other atomic components. Consequently, atomic components represent the smallest executable entities of a program and establish the basis for a tunable architecture. The notion of an atomic component does not imply anything about locking or atomicity in the conventional concurrency sense.

We assume that an atomic component is implemented as a method in an object-oriented program. Atomic components can be marked as *replicable* if the associated method is stateless (i.e., the result of a method execution does not depend on previous runs). The runtime system is allowed to create several instances of replicable component that execute in parallel.

In TADL, the declaration of an atomic component is simply the name of a method followed by an optional `replicable` attribute:

```
MyMethodName [replicable]
```

TADL establishes the association between a method and the atomic component declaration using the method's name.

10.5.1.2 Connectors

Connectors are operators composing atomic components or other connectors. Thus, a tunable architecture has a tree structure with connectors as nodes and atomic components as leaves. In contrast to the definition of connectors in architecture description languages [16], where a connector handles the communication between two components, a TADL connector defines a set of components including the interactions among those components.

TADL provides five connectors types: (1) *Sequential composition* encloses child items that must not execute concurrently. (2) *Tunable alternative* expresses an exclusive alternative of two or more components. (3) *Tunable fork-join* introduces task parallelism. That is, all enclosed components run in parallel (fork) followed by an implicit barrier (join). (4) *Tunable producer-consumer* provides a technique to pass data from one component to another. (5) *Tunable pipeline* introduces pipeline parallelism. Each enclosed component represents a pipeline stage.

A connector encloses a set of child items that can be either atomic components or other connectors. Every child item enclosed by a connector can have an associated input and output component for providing and retrieving data. Similar to atomic components, we assume that input and output components are implemented as methods in an object-oriented program.

10.5.1.3 Runtime System and Backend

The backend of tunable architectures and TADL consists of the *tunable architecture library (TALib)* and the *TADL compiler*.

The TALib has modules implementing every connector type mentioned earlier. For example, the Tunable Pipeline connector is implemented in the TALib's Tunable Pipeline module. In addition, each module exposes a set of predefined tuning parameters. There are default parameter values ranges, but these can be modified by the programmer.

The *TADL compiler* generates source code with *tuning wrappers* from the TADL script. A tuning wrapper is a class that initializes a particular TALib module and implements access methods to that module. For example, the compiler translates a tunable pipeline connector into a tuning wrapper that handles the access to the TALib's tunable pipeline module.

The TADL compiler also inserts instrumentation annotations into the atomic components and thus creates a runtime environment for the auto-tuner. We use *Atune-IL*, which is a `pragma`-based tuning language [25] described later. To provide tuning instructions to the auto-tuner, the tuning wrappers also contain code instrumentations specifying parameter value ranges, measuring points, and parameter context information.

10.5.1.4 A Tunable Architecture Example

The following example is taken from [26] and describes the design of a parallel video processing application. It applies a sequence of filters to frames of a video stream.

The initial program implementation contains four filter methods (`Crop()`, `OilPaint()`, `Resize()`, and `Sharpen()`). These methods implement atomic components. Two more methods, `GetVideo()` and `Consume()` implement input and output components.

The tunable architecture description is shown in Listing 10.1. The tunable pipeline arranges the enclosed atomic components in a pipelined fashion. `Source` and `sink` identify the input and output components. As the processing of a particular frame is stateless, all child items are marked replicable. Thus, stages that are bottlenecks can be replicated.

Listing 10.1: Architecture Description of the Parallel Video Processing Application

```
TunablePipeline MyVideoProcessing
    [ source : LoadVideo ;
    sink : ConsumeVideo]  {
    Crop[ replicable ],
    OilPaint[ replicable ],
    Resize[ replicable ],
    Sharpen[ replicable ]
}
```

Source: Schaefer, C.A et al., Engineering parallel applications with tunable architectures, in *Proceeding of the 32nd ACM/IEEE International Conference on Software Engineering, Volume I, ICSE'10*, Cape Town, South Africa, pp. 405–414, 2010, ACM, New York.

As the example shows, tunable architectures allow the expression of performance-relevant architecture variations. The programmer needs to provide only the TADL architecture description and the atomic components. The rest is completely automatic, including making the software tunable, inserting measuring points, and running the tuning process.

10.5.1.4.1 Performance Improvements with Tunable Architectures   Figure 10.4 shows performance results with tunable architectures on six applications. The applications were executed on an eight-core machine ($2\times$ Intel Xeon E5320 Quad-core, 1.86 GHz, 8 GB RAM, .Net).

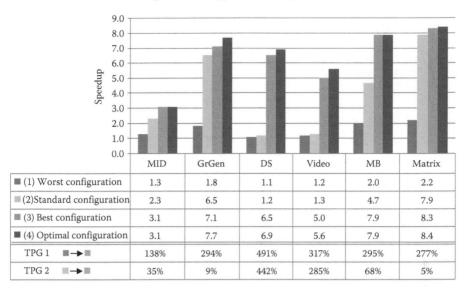

	MID	GrGen	DS	Video	MB	Matrix
■ (1) Worst configuration	1.3	1.8	1.1	1.2	2.0	2.2
▨ (2)Standard configuration	2.3	6.5	1.2	1.3	4.7	7.9
■ (3) Best configuration	3.1	7.1	6.5	5.0	7.9	8.3
■ (4) Optimal configuration	3.1	7.7	6.9	5.6	7.9	8.4
TPG 1 ■→■	138%	294%	491%	317%	295%	277%
TPG 2 ▨→■	35%	9%	442%	285%	68%	5%

FIGURE 10.4: Application performance improvements.

MID is the commercial biological data analysis application that has been parallelized as described earlier in Section 10.2. The performance numbers are obtained for an input file of 1 GB that contains mass spectrogram data. GrGen is a graph rewriting system. The benchmark simulates gene expression as a graph rewriting task on *E. coli* DNA with over nine million graph nodes. DS is a parallel desktop search application that generates an index for text files. The input consists of over 10,000 ASCII text files with sizes between 6 kB and 613 kB. The video processing application applies four filters in parallel on the picture stream of an AVI file, as discussed earlier. The input consists of 180 frames at a resolution of 800 × 600 pixels. MB computes the Mandelbrot set and Matrix is the multiplication of two 1024 × 1024 matrices.

The table in Figure 10.4 shows for each application the parallel speedup with the worst configuration (1), with a default configuration (2), with the best configuration found by the auto-tuner (3), and the optimum configuration (4) obtained by exhaustive search. The default configuration is what a competent developer might guess. The tuning performance gain between the worst and the best found configuration (TPG1) shows that auto-tuning produces significant speedup gains, from 138% to almost 500%. The performance gain between the default configuration and the best found configuration (TPG2) is smaller. As can be seen, the developer's guess is far off the optimum in two cases, and moderately off in two others. The auto-tuner always finds a configuration that is close to the optimum.

It is also interesting to note that the auto-tuner may choose non-obvious configurations. In the video application, the final stage is the slowest one. In the optimal configuration, this stage is replicated eight times. Since the other stages are quite fast, they are fused with the replicates of the final stage. The result is a pipeline that is replicated eight times, but each replica runs sequentially. Thus, there is no buffering or synchronization except at the beginning and the end of the pipelines. DS also has a non-obvious configuration. Apparently, it is precisely in those non-obvious situations where the auto-tuner works best.

10.5.2 CO^2P^3S

CO^2P^3S (*correct object-oriented pattern-based parallel programming system*) [15] combines object-oriented programming, design patterns, and frameworks for parallel programming. The approach only supports the development of new parallel programs.

CO^2P^3S has a graphical interface. The user chooses a parallel pattern for the entire program structure. The system generates the appropriate code including instructions for communication and synchronization. The generated code provides a predefined structure with placeholders (called *hook methods*) for sequential code. The user has to implement these hook methods by inheriting the corresponding framework classes and overwriting the methods. The user can thus choose to work on the intermediate code layer or the native code layer.

The approach uses parameterized pattern templates to influence the code generation process. The programmer can configure the selected pattern and adapt it for a specific use. For example, the parameters for a two-dimensional regular mesh template include the topology, the number of neighboring elements, and the synchronization level [15]. Although the selection of an appropriate pattern variant can influence program performance, the template parameters do not necessarily represent tuning parameters, as they do not primary address performance bottlenecks relevant for parallelism. Moreover, the parameters cannot be directly adjusted after code generation.

The approach aims to simplify the creation of a parallel application and consists of the following five steps [15]:

1. Identify one or more parallel design patterns that can contribute to a solution and pick the corresponding design pattern templates.

2. Provide application-specific parameters for the design pattern templates to generate collaborating framework code for the selected pattern templates.

3. Provide application-specific sequential hook methods and other nonparallel code to build a parallel application.

4. Monitor the parallel performance of the application and if it is not satisfactory, modify the generated parallel structure code at the intermediate code layer.

5. Re-monitor the parallel performance of the application and if it is still not acceptable, modify the actual implementation to fit the specialized target architecture at the native code layer.

Even though related to tunable architectures, CO^2P^3S focuses on the generation of pattern-related framework code rather than on creating a tunable program architecture. The work so far lacks an auto-tuner.

10.5.3 Comparison

Both approaches exploit parallelism at the software architecture level. Tunable architectures consist of a description language to describe parallel architectures. In contrast to CO^2P^3S, the association between program code and architecture description is part of the design process. This makes tunable architectures applicable for new programs as well as for existing programs. Architecture adaptation and automatic performance optimization represent an integral part of the tunable architectures approach. CO^2P^3S employs the generation of framework code that uses calls to hook methods to integrate application-specific code. The framework supplies the application structure and the programmer writes application-specific utility routines. The configuration of the program and its patterns happens before code generation, which results in a hard-coded variant. Manual performance tuning is required to optimize the patterns after implementation.

10.6 Implementation with Tuning Instrumentation Languages

In the previous section, we have defined the characteristics of a tunable parallel application. To be tunable, a parallel application must expose variables that influence its performance. Therefore, the program must be adaptable and change its behavior according to the variables' values. However, a tunable application is not auto-tunable by default, as an auto-tuner does not know the performance-relevant variables and how to retrieve performance feedback. To develop a parallel application with auto-tuning support, tuning-relevant spots have to be marked for the auto-tuner.

In principle, tuning information can be added to a program in two ways. As a straightforward technique, tuning parameters, measuring points, and other tuning-relevant functionality can be integrated within the program's code. To adjust a tuning parameter, one can pass the values, for instance, via command line (using off-line tuning) or via library calls (using online tuning). This approach is useful for smaller programs that are intended to run on one particular hardware platform. A drawback, however, is that tuning-relevant code is mixed with the other program code, so changes might complicate code reengineering.

A more flexible approach to specify tuning information is using domain-specific scripting languages or instrumentation languages. These languages allow the definition of tuning-relevant instructions, such that tuning meta data is separated from program code. This approach not only improves maintainability, but also decouples the auto-tuner from the program. Auto-tuners thus become easier to use on a wider range of general applications.

A number of projects designed tuning languages. The languages differ greatly in generality, applicability to parallel programs, efficiency, and the effort to use them. We present next the work we consider most relevant from a software engineering perspective. Many approaches combine tuning languages with an optimization component that works like an auto-tuner, even though the auto-tuner is often not a separate entity. Auto-tuning systems will be discussed in Section 10.7.

10.6.1   Atune-IL

Atune-IL is used in our auto-tuning system. It targets parallel applications in general, not just numerical programs, and works with empirical auto-tuners [25]. Atune-IL uses `pragma` directives (such as `#pragma` for C++ and C# or `/*@` for Java) that are inserted into regular code. Atune-IL provides constructs for declaring tuning parameters, permutation regions (to allow an auto-tuner to reorder source code statements), and measuring points. Additional statements describe software architecture meta-information that helps an empirical auto-tuner reduce the search space *before* employing search algorithms. Therefore, Atune-IL is able to capture the structure of an application along with characteristics important for tuning, such as parallel sections and dependencies between tuning parameters.

Atune-IL uses tuning blocks to define scopes and visibility of tuning parameters and to mark program sections that can be optimized independently. A developer can mark two code sections to be independent if they cannot be executed concurrently in any of the application's execution paths. Tuning blocks can be nested. Global tuning parameters and measuring points are automatically bound to an implicit root tuning block that wraps the entire program. Thus, each parameter and each measuring point belongs to a tuning block (implicitly or explicitly). Each independent block structure represents a *tuning entity*.

Tunable parameters can be defined with the `setvar` statement, which means that the value of variable preceded by such a statement can be set by an auto-tuner. The number of threads to create is an example of such a variable. A measuring point is defined by inserting the `gauge` statement. Two such statements with the same identifier in the same block define two measuring points whose difference will be used by the auto-tuner. Tuning blocks are defined by `startblock` and `endblock`. Listing 10.2 shows an example of an instrumented program that searches strings in a text, stores them in an array, and sorts the array using parallel sorting algorithms. Finally, the program counts the total number of characters in the array. The tuning block `sortBlock` is logically nested within `fillBlock`, while `countBlock` is considered to be independent from the other tuning blocks. Thus, the parameters `sortAlgo` and `depth` in `sortBlock` have to be tuned together with parameter `stringSearch` in `fillBlock`, as the order of the array elements influences sorting. This results in two separate tuning entities that can be tuned one after the other. The sizes of the search spaces are $dom(fillOrder) \times dom(sortAlgo) \times dom(depth)$ and $dom(numThreads)$, respectively. In addition, the dependency of *depth* on *sortAlgo* can be used to prune the search space, because invalid combinations are ignored.

Listing 10.2: C# Program Instrumented with Atune-IL

```
List<string> words = new List<string>(3);
void main() {
    #pragma atune startblock fillBlock
    #pragma atune gauge mySortExecTime

    string text = "Auto-tuning has nothing to do with tuning cars."
```

```
    #pragma atune startpermutation fillOrder
    #pragma atune nextelem
    words.Add(text.Find("cars"));
    #pragma atune nextelem
    words.Add(text.Find("do"));
    #pragma atune nextelem
    words.Add(text.Find("Auto−tuning"));
    #pragma atune endpermutation

    sortParallel(words);

    #pragma atune gauge mySortExecTime
    #pragma atune endblock fillBlock

    countWords(words)
}

// Sorts string array
void sortParallel(List<string> words) {
    #pragma atune startblock sortBlock inside fillBlock

    IParallelSortingAlgorithm sortAlgo = new ParallelQuickSort();
    int depth = 1;
    #pragma atune setvar sortAlgo type generic
        values "new ParallelMergeSort(depth)","new ParallelQuickSort()"
        scale nominal
    #pragma atune setvar depth type int
        values 1−4 scale ordinal
        depends sortAlgo='new ParallelMergeSort(depth)'

    sortAlgo.Run(words);

    #pragma atune endblock sortBlock
}

// Counts total characters of string array
int countCharacters(List<string> words) {
    #pragma atune startblock countBlock
    #pragma atune gauge myCountExecTime
    int numThreads = 2;
    #pragma atune setvar numThreads type int
        values 2−8 scale ordinal

    int total = countParallel(words, numThreads);
```

```
#pragma atune gauge myCountExecTime
return total;
#pragma atune endblock countBlock
}
```

Source: Schaefer, C.A et al., Atune-IL: An instrumentation language for auto-tuning parallel applications, in *Proceeding of the 15th International Euro-Par Conference on Parallel Processing.* Volume 5704/2009 of *LNCS*, Delft, the Netherlands, pp. 9-20, Springer, Berlin/Heidelberg, Germany, January 2009.

Atune-IL is one of the first tuning instrumentation languages to provide a variety of constructs that support auto-tuners in the reduction of the search space.

10.6.2 X-Language

The *X-Language* [4] provides constructs to declare source code transformations. The software developer specifies transformations on particular parts of the source code (such as a loop). X-Language uses #pragma directives to annotate C/C++ source code. The X-Language interpreter applies the specified code transformations before the program is compiled.

The X-Language offers a compact representation of several program variants and focuses on fine-grained loop optimizations. Listing 10.3 shows a simple example of how to annotate a loop to automatically unroll it with a factor of 4. Listing 10.4 shows the code resulting from the transformation.

Listing 10.3: Annotated Source Code with X-Language

```
#pragma xlang name l1
for (i=0; i<256; i++) {
    s = s + a[i];
}
#pragma xlang transform unroll l1 4
```

Listing 10.4: Source Code after X-Language Transformation

```
sum=0;
#pragma xlang name l1
for (i=0; i<256; i+=4) {
    s = s + a[i];
    s = s + a[i+1];
    s = s + a[i+2];
    s = s + a[i+3];
}
```

Source: Donadio, S. et al., A language for the compact representation of multiple program versions, in *Proceedings of the 18th International Workshop on languages and Compilers for Parallel Computing*, Volume 4339 of *LNCS*, Montreal, Quebec, Canada, pp. 136–151, 2006.

Custom code transformations can be defined with pattern matching rewrite rules. The X-Language does not include an optimization mechanism.

10.6.3 POET

POET (parameterized optimizing for empirical tuning [30]) is a domain-specific language for source code transformations. POET uses an XML-like script to specify transformations, parameters, and source code fragments. A source code generator produces a compilable program based on such a script.

Compared to the X-Language, a POET script contains the complete code fragments of the original application that are relevant for tuning. POET also allows the generic definition of transformations that can be reused in other programs. POET focuses on small numerical applications solving one particular problem, and does not provide explicit constructs to support parallelism. It works at the detailed statement level rather than the architecture level and is quite difficult to use. Specifying the loop unrolling in the previous example requires 30 lines of POET code.

10.6.4 Orio

The *Orio* [9] tuning system consists of an annotation language for program variant specification and an empirical tuning component to find the best-performing variant.

Orio parses annotated source code (limited to C code), generates new source code to create program variants, and searches for the variant with the shortest execution time. Although this process is straightforward, Orio's annotation language is complex. For example, instrumenting a single loop in the program's source (e.g., for automatic loop unrolling) requires a large external script and multiple annotations. Moreover, parts of the source code have to be redundantly embedded within the annotation statements.

Orio uses separate specifications for tuning parameters, while the source code annotations contain the actual transformation instructions. This technique allows the reuse of parameter declarations and keeps the source code cleaner. However, the annotations entirely replicate the source code in order to place the transformation instructions. This results in code duplication that is not suitable for large applications.

Orio supports automatic code transformation at loop level. Its tuning component uses two heuristics: the *Nelder–Mead simplex* method and *simulated annealing*. However, the experimental setup presented in [9] demonstrates applicability only on small numerical programs.

10.6.5 Comparison

The tuning instrumentation languages mainly differ in their level of granularity and in the targeted problem domain. Orio and POET are suitable for numerical applications and perform typical code transformations on loop level required in that domain. From a software engineering perspective, they have the drawback of requiring code duplication. This limits their applicability for larger projects and makes maintenance and debugging difficult. The X-Language also targets scientific programs and loop level transformations. It permits custom code transformations, which make it more flexible. Atune-IL is among the first languages to focus on general-purpose parallel programs. Its language constructs work beyond loop transformations and it allows the declaration of tuning parameters, measuring points, and scopes for tuning entities. Atune-IL does not support automatic loop unrolling, although it permits handwritten alternatives.

10.7   Performance Optimization

In the previous sections, we have discussed how to design tunable parallel applications and how to specify tuning instructions within the program. We now consider the tuning process itself. At this point, we assume that a program exposes relevant performance parameters and measuring points to an empirical auto-tuner, which tries out different configurations to find the best performance.

10.7.1   Auto-Tuning Cycle

Figure 10.5 shows the typical auto-tuning cycle [25].

Initially, information about tuning parameters is passed to the auto-tuner (0). Then, the auto-tuner calculates an initial parameter configuration (1). The program is configured with the calculated parameter values (2). The auto-tuner executes the new program variant and monitors performance. After termination, performance values are stored to direct the search in the next cycle (3).

The tuning cycle in steps 1–3 is repeated until some predefined condition is met. The condition depends on the method employed in the optimization step. We call

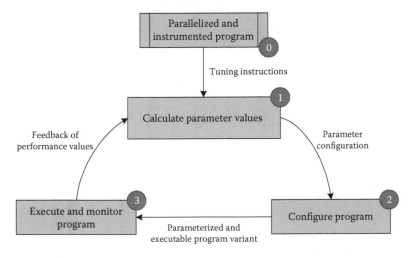

FIGURE 10.5:   Basic auto-tuning cycle.

one cycle of steps 1–3 a *tuning iteration*. With each tuning iteration, the auto-tuner samples the search space and maps a parameter configuration to a performance value.

10.7.2 Search Techniques

Empirical tuning employs meta-heuristics. A meta-heuristic algorithm solves a combinatorial optimization problem by approximation. All elements of the search space are potential solutions. The algorithm tests solutions until some convergence criterion holds. However, it is not guaranteed that meta-heuristics find an optimum.

As defined earlier in Section 10.3.1, each parameter configuration C within the search space S represents a valid solution to configure the program $\wp$. We determine the quality of a parameter configuration C using its performance value $\ell_{\wp,C}$. The parameter configuration yielding the best performance value is interpreted as the optimal configuration of $\wp$.

We evaluate the quality of a meta-heuristic by two metrics. The first is the mean error rate, that is, how good the results are. In addition, we are interested in how quick the algorithm converges, that is, how many parameter configurations have to be tested until the algorithm terminates.

A comprehensive discussion of search algorithms is far beyond the scope of this chapter. We therefore present an overview and refer to [17,20] for details and other algorithms.

Random sampling: This approach randomly selects a set of parameter configurations out of the search space and tests them. The parameter configuration yielding the best performance value is returned. Random sampling can provide useful results when the search space is small or when most of the configurations have similar performance values.

Local search: This approach focuses on a local region around an initial configuration. The search only considers parameter configurations that exist within a well-defined neighborhood of the start configuration.

More specifically, well-known local search algorithms are, for example:

- *Hill climbing*: This widely used algorithm randomly selects an initial configuration and follows the steepest assent in the neighborhood. A major drawback is that the algorithm can get stuck in local optima.

- *Simulated annealing*: The inspiration for this approach comes from the processes of annealing in metallurgy. Controlled heating and cooling of a material results in a stable crystal structure with reduced defects. The process must be performed slowly in order to form an optimal uniform structure of the material. By analogy with this physical process, simulated annealing does not always continue with the best configurations so far, but temporarily chooses a poorer one. The probability for this choice is governed by a global parameter T (the temperature). T is gradually decreased for each tuning iteration. This method saves the algorithm from getting stuck in local optima. Simulated annealing thus improves hill climbing. It allows a temporary decline of the tested configuration's quality, but gains the advantage of finding the global optimum in the long run.

- *Tabu search*: This approach performs an iterative search as well, but keeps a *tabu list* that contains parameter configurations that are temporarily excluded from testing. This method avoids a circular exploration of the search space. A commonly used tabu strategy adds the current configuration's complement to the list.

Global search: Global search algorithms start with several initial parameter configurations. Global search algorithms differ from local searchers mainly in the strategy defining how new configurations are selected. Well-known approaches are swarm optimization [13] and genetic algorithms [6].

10.7.3   Auto-Tuning Systems

This section provides an overview of complete auto-tuning systems.

10.7.3.1   Atune

Atune [24,25] is our off-line empirical auto-tuner. It works with programs instrumented with Atune-IL (see Section 10.6.1). In combination with TADL and its parallel patterns, Atune supports the automatic optimization of large parallel applications on an architectural level. The auto-tuner is general-purpose, that is, not tied to a specific problem domain.

Atune follows a multistage optimization process:

1. *Create tuning entities*: First, Atune analyzes the program structure using Atune-IL's tuning blocks and creates tuning entities. The result of this step are one or more search space partitions that will be handled independently. The partitions are automatically tuned one after the other.

2. *Analyze tuning entities*: For each tuning entity, Atune extracts the tuning instructions, such as parameter declarations, and measuring points. Furthermore, it classifies the tuning parameters according to their scale (nominal or ordinal). This differentiation is important for the tuning process, as common search techniques require an ordinal scale. Nominally scaled parameters, for instance for expressing algorithm alternatives, must be handled differently. Atune first samples the nominal parameters randomly and then continues with the best choice for them into the next optimization phase.

   Atune also gathers information about the patterns present and applies suitable heuristics in this case. For example, if the tuning entity consists of the pipeline pattern, it will use a special search procedure that balances the pipeline stages by replicating and fusing stages.

3. *Search*: Each tuning entity that still contains unconfigured parameters will be auto-tuned using search algorithms like the ones described in the previous section.

The multistage optimization process reduces the search space, so Atune is able to handle complex parallel applications consisting of several (nested) parallel sections exposing a large number of tuning parameters.

10.7.3.2 ATLAS/AEOS

ATLAS (automatically tuned linear algebra software) [29] combines a library with a tuning tool called *AEOS (automated empirical optimization of software)* to find the fastest implementation of a linear algebra operation on a specific hardware platform.

The tool generates high-performance libraries for a particular hardware platform using domain-specific knowledge about the problem to solve. AEOS performs the source code optimizations in an iterative off-line tuning process, while ATLAS represents the optimized library.

The authors of AEOS/ATLAS also developed a model-based tuning approach. Instead of feeding the performance results back into the optimization process to gradually adjust the parameter values, a manually created analytical model is used to directly calculate the best parameter configuration.

An enhancement of AEOS/ATLAS is described in [5]. The approach combines model-based and empirical tuning. The idea is to reduce the initial search space using an analytical model, before the search starts. The models are created using machine learning approaches. If limited to the numerical programs targeted in ATLAS, the approach offers promising results. Furthermore, this work is one of the few approaches that specifically targets the reduction of the search space before applying any search algorithm.

Although AEOS/ATLAS and other tuning approaches for numerical application (such as FFTW [7], FIBER [12], and SPIRAL [23]) form the background of today's auto-tuning concepts, they focus on small application domains and numerical problems.

10.7.3.3 Active Harmony

Active Harmony [28] provides a runtime system that optimizes programs regarding network and system resources. The approach focuses on exchangeable libraries rather than on particular tuning parameters. A library developer can create several variants of a particular library using a predefined API to let the auto-tuner decide which implementation is best for performance. All exchangeable libraries must use the Active Harmony API.

Active Harmony employs an empirical off-line auto-tuner that improves the performance of all exchangeable libraries in the program. The initial version of Active Harmony applies the Downhill Simplex method to explore the search space. Further work [27] improves the search algorithm and discusses adaptations to match the tuning requirements in a more precise way.

Unlike other approaches, Active Harmony abstracts from fine-grained optimizations and performs performance-relevant adaptations on higher abstraction levels. Moreover, Active Harmony is not limited to particular application domains. However, the approach is not explicitly focused on shared-memory parallel programs.

10.7.3.4 Model-Based Systems

In recent years, model-based approaches evolved in addition to empirical tuning approaches. As it is very difficult to develop general models that allow conclusions about the overall program performance, all approaches deal with performance modeling and analysis of parallel patterns [2,18]. The key idea is to exploit the knowledge about behavior and structure of parallel patterns to identify performance bottlenecks, which in turn represent potential tuning parameters. Doing so, the models can help predict the performance of relevant parallel parts of the program. Some of the approaches (such as [18]) combine the analytical models with online tuning.

The experimental results of the research projects show that suitable models are able to precisely predict the performance of a particular parallel pattern (e.g., Master/ Worker [1] or Pipeline [3]). For this purpose, the models must fit exactly to the application domain and the hardware platform. Some approaches even include parameters related to programming models (such as MPI [19] or PVM) into the models. Precision can be improved, but the approach significantly narrows down the applicability of models with respect to different platforms, program types, and parallel problems. If the program needs to be modified or ported to another platform, the models must be adjusted or entirely recreated.

Unfortunately, model-based approaches usually do not consider applications consisting of several parallel sections that could each implement a different parallel pattern. In practice, empirical auto-tuning still plays an important role, and model-based approaches can help reduce the search space.

10.7.4 Comparison

The aforementioned auto-tuning approaches implement different techniques to find the best possible parameter configuration of a program. The systems mainly differ in their assumptions about the application domains and in the targeted software abstraction levels. While AEOS/ATLAS focuses on optimizing linear algebra operations, Active Harmony supports optimized libraries. Atune provides tuning techniques that specifically target large parallel applications to enable optimization on a software architecture level. Furthermore, Atune is not restricted to particular problem domain. An enhanced version of AEOS/ATLAS [5] as well as Atune reduce the search space prior to the actual search. While AEOS/ATLAS uses domain-specific models to predefine parameters of the algebraic library, Atune gathers context information about the program structure using the tuning instrumentation language Atune-IL.

10.8 Conclusion and Outlook

The general auto-tuning approaches presented in this chapter have shown promising results for both the application performance optimization as well as for the simplification of software development. This is just the beginning, and there are plenty of opportunities for further work. For example, more parallel patterns need to be studied with respect to tuning heuristics. It is important to understand how these patterns are used for tuning. A good starting point might be pattern-based performance diagnosis [14]. Another promising direction is to improve the interaction between compilers and auto-tuners and implicitly introduce auto-tuning [11]. Yet another largely neglected area is simultaneous tuning of several parallel programs [10]: While most of the current research concentrates on optimizing program performance for just one program in isolation, in practice the performance is influenced by other processes competing for available resources on the same machine. A global auto-tuner working at the level of the operating system is key to finding optimum performance in such complex situations.

References

1. E. Csar, J.G. Mesa, J. Sorribes, and E. Luque. Modeling master/worker applications in POETRIES. In *Proceedings of the ninth International Workshop on High-Level Parallel Programming Models and Supportive Environments*, Santa Fe, NM, pp. 22–30, April 2004.

2. E. Csar, A. Moreno, J. Sorribes, and E. Luque. Modeling master/worker applications for automatic performance tuning. *Parallel Computing*, 32(7–8):568–589, 2006. Algorithmic Skeletons.

3. E. Csar, J. Sorribes, and E. Luque. Modeling Pipeline Applications in POETRIES. In *Proceedings of the Eighth International Euro-Par Conference on Parallel Processing*, Lisbon, Portugal, pp. 83–92, 2005.

4. S. Donadio, J. Brodman, T. Roeder, K. Yotov, D. Barthou, A. Cohen, M. Garzaran, D. Padua, and K. Pingali. A language for the compact representation of multiple program versions. In *Proceedings of the 18th International Workshop on Languages and Compilers for Parallel Computing*, vol. 4339 of *LNCS*, Montreal, Quebec, Canada, pp. 136–151, 2006.

5. A. Epshteyn, M. Jess Garzaran, G. DeJong, D. Padua, G. Ren, X. Li, K. Yotov, and K. Pingali. Analytic models and empirical search: A hybrid approach to code optimization. In *Proceedings of the Workshop on Languages and Compilers for Parallel Computing*, vol. 4339/2006, New Orleans, LA, pp. 259–273, 2006.

6. L.J. Fogel, A.J. Owens, and M.J. Walsh. *Artificial Intelligence through Simulated Evolution*. John Wiley & Sons, New York, 1966.

7. M. Frigo and S.G. Johnson. FFTW: An adaptive software architecture for the FFT. In *Proceedings of the International Conference on Acoustics, Speech and Signal Processing*, vol. 3, Seattle, WA, pp. 1381–1384, May 1998.

8. E. Gamma, R. Helm, R. Johnson, and J. Vlissides. *Design Patterns: Elements of Reusable Object-Oriented Software*. Addison-Wesley Professional, Boston, MA, 1995.

9. A. Hartono and S. Ponnuswamy. Annotation-based empirical performance tuning using orio. In *Proceedings of the 23rd IEEE International Parallel and Distributed Processing Symposium (IPDPS)*, Rome, Italy, May 2009.

10. T. Karcher and V. Pankratius. Run-time automatic performance tuning for multicore applications. In *Proceedings Euro-Par2011*, LNCS 6853, Bordeaux, France, September 2011.

11. T. Karcher, C. Schaefer, and V. Pankratius. Auto-tuning support for manycore applications: Perspectives for operating systems and compilers. *ACM SIGOPS Operating Systems Review*, 43(2):96–97, 2009.

12. T. Katagiri, K. Kise, H. Honda, and T. Yuba. FIBER: A generalized framework for auto-tuning software. In *Proceedings of the International Symposium on High Performance Computing*, vol. 2858/2003, Tokyo-Odaiba, Japan, pp. 146–159, 2003.

13. J. Kennedy and R. Eberhart. Particle swarm optimization. In *Proceedings of the IEEE International Conference on Neural Networks*, Perth, Western Australia, Australia, pp. 1942–1948, 1995.

14. L. Li and A.D. Malony. Knowledge engineering for automatic parallel performance diagnosis. *Concurrency and Computation: Practice and Experience*, 19:1497–1515, 2007.

15. S. MacDonald, J. Anvik, S. Bromling, J. Schaeffer, D. Szafron, and K. Tan. From patterns to frameworks to parallel programs. *Journal of Parallel and Distributed Computing*, 28(12):1663–1683, 2002.

16. N. Medvidovic and R.N. Taylor. A classification and comparison framework for software architecture description languages. *IEEE Transactions on Software Engineering*, 26(1):70–93, 2000.

17. Z. Michalewicz and D.B. Fogel. *How to Solve It: Modern Heuristics*. Springer Verlag, Berlin, Germany, 2000.

18. A. Morajko, E. César, P. Caymes-Scutari, T. Margalef, J. Sorribes, and E. Luque. Automatic tuning of master/worker applications. In *Proceedings of the Eighth International Euro-Par Conference on Parallel Processing*, Paderborn, Germany, pp. 95–103, 2005.

19. MPI Forum. *The Message Passing Interface (MPI) Standard*. MPI Forum, 2009. http://www.mcs.anl.gov/research/projects/mpi/ (last accessed August 2011.)

20. J.A. Nelder and R. Mead. A simplex method for function minimization. *The Computer Journal*, 7(4):308–313, January 1965.

21. V. Pankratius, A. Jannesari, and W.F. Tichy. Parallelizing BZip2. A case study in multicore software engineering. *IEEE Software*, 26(6):70–77, 2009.

22. V. Pankratius, C.A. Schaefer, A. Jannesari, and W.F. Tichy. Software engineering for multicore systems: An experience report. In *Proceedings of the International Workshop on Multicore Software Engineering*, Leipzig, Germany, pp. 53–60, 2008. ACM, New York.

23. M. Püschel, J.M.F. Moura, J.R. Johnson, D. Padua, M.M. Veloso, B.W. Singer, J. Xiong, F. Franchetti, A. Gacic, Y. Voronenko, K. Chen, R.W. Johnson, and N. Rizzolo. SPIRAL: Code generation for DSP transforms. *Proceedings of the IEEE*, 93(2):232–275, February 2005.

24. C.A. Schaefer. Reducing search space of auto-tuners using parallel patterns. In *Proceedings of the 2nd ICSE Workshop on Multicore Software Engineering*, Vancouver, British Columbia, Canada, pp. 17–24, 2009. IEEE Computer Society, Washington, DC.

25. C.A. Schaefer, V. Pankratius, and W.F. Tichy. Atune-IL: An instrumentation language for auto-tuning parallel applications. In *Proceedings of the 15th International Euro-Par Conference on Parallel Processing*, vol. 5704/2009 of *LNCS*, Delft, the Netherlands, pp. 9–20. Springer Berlin/Heidelberg, Germany, January 2009.

26. C.A. Schaefer, V. Pankratius, and W.F. Tichy. Engineering parallel applications with tunable architectures. In *Proceedings of the 32nd ACM/IEEE International Conference on Software Engineering, ICSE '10*, vol. 1, Cape Town, South Africa, pp. 405–414, 2010. ACM, New York.

27. V. Tabatabaee, A. Tiwari, and J.K. Hollingsworth. Parallel parameter tuning for applications with performance variability. In *Proceedings of the ACM/IEEE Supercomputing Conference*, Heidelberg, Germany, pp. 57–57, November 2005.

28. C. Tapus, I-H. Chung, and J.K. Hollingsworth. Active harmony: Towards automated performance tuning. In *Proceedings of the ACM/IEEE Supercomputing Conference*, New York, November 2002.

29. R.C. Whaley, A. Petitet, and J.J. Dongarra. Automated empirical optimizations of software and the ATLAS Project. *Journal of Parallel Computing*, 27:3–35, January 2001.

30. Q. Yi, K. Seymour, H. You, R. Vuduc, and D. Quinlan. POET: Parameterized optimizations for empirical tuning. In *Proceedings of International Parallel and Distributed Processing Symposium*, Long Beach, CA, pp. 1–8, March 2007.

31. C.A. Schaefer, Automatische Performanzoptimierung Paralleler Architekturen (PhD thesis in German), Karlsruhe Institute of Technology, 2010.

Chapter 11

Transactional Memory

Tim Harris

Contents

11.1   Introduction

Many of the challenges in building shared-memory data structures stem from needing to update several memory locations at once—e.g., updating four pointers to insert an item into a doubly linked list. Transactional memory (TM) provides a mechanism for grouping together this kind of series of operations, with the effect that either all of them appear to execute, or none of them does. As with database transactions, TM lets the programmer focus on the changes that need to be made to data, rather than dealing with the low-level details of exactly which locks to acquire, or how to prevent deadlock.

In this chapter, we look at how TM can be used by a programmer, at the ways it can be implemented in hardware (HTM) and software (STM), and at how higher-level constructs can be built over it in a programming language. As a running example,

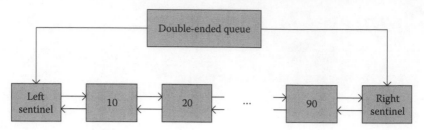

FIGURE 11.1:   A double-ended queue built over a doubly linked list with sentinel nodes.

we use an implementation of a double-ended queue (deque). Figure 11.1 shows the structure of the deque when it contains the numbers 10, 20, . . . , 90. The queue object refers to left and right sentinel nodes, with the queue's contents strung between the sentinels in a doubly linked list.

It is easy to update a deque in a single-threaded program. Figure 11.2 shows a fragment of the implementation in pseudo-code, focusing on the fields containing links from the queue to the sentinels, and the implementation of the `pushLeft` operation.

How could a scalable version of this simple data structure be built using locks? One option is to use a single lock to protect the entire deque. This would be straightforward, and a basic spin-lock would add only one or two instructions to each operation on the deque. However, the lock would serialize operations unnecessarily—e.g., it would prevent one thread from removing 10 from the left in parallel with another thread removing 90 from the right.

Scalable designs are much more complicated. We might try to use two locks, arranging that `pushLeft`/`popLeft` acquires one lock and `pushRight`/ `popRight` acquires the other. That will work when the deque contains several

```
Class Q {
  QElem leftSentinel;
  QElem rightSentinel;

  void pushLeft(int item) {
    QElem e = new QElem(item);
    e.right = this.leftSentinel.right;
    e.left = this.leftSentinel;
    this.leftSentinel.right.left = e;
    this.leftSentinel.right = e;
  }

  ...
}
```

FIGURE 11.2:   Sequential implementation of 'pushLeft'.

```
Class Q {
    QElem leftSentinel;
    QElem rightSentinel;

    void pushLeft(int item) {
        QElem e = new QElem(item);
        e.left = this.leftSentinel;
        do {
            e.right = LT(&this.leftSentinel.right);
            ST(&LT(&this.leftSentinel.right).left, e);
            ST(&this.leftSentinel.right, e);
        } while (!COMMIT());
    }
    ...
}
```

FIGURE 11.3: Implementation of 'pushLeft' with explicit transactional operations.

elements, but what about the case when the deque is almost empty? It is difficult to avoid deadlock when a thread needs to acquire both locks.

In this particular example, for a deque, several good scalable solutions are known. However, designing deques is still the subject of research papers, and it is difficult and time-consuming to produce a good design. TM aims to provide an alternative abstraction for building this kind of shared-memory data structure, rather than requiring ingenious design work to be repeated for each new case.

Figure 11.3 shows how the pushLeft operation could be implemented using Herlihy and Moss' HTM (we look at this HTM in detail in Section 11.3). The code is essentially the same as the sequential version, except that the related operations are grouped together in the do...while loop, and the individual reads and writes to shared memory are made using two new operations: LT ("load transactional") and ST ("store transactional").

This looping structure is typical when using TM; all of the implementations we describe rely on optimistic concurrency control, meaning that a thread speculatively attempts a transaction, checking whether or not its speculative work conflicts with any concurrent transactions. At the end of a transaction, the speculative work can be made permanent if no conflicts occurred. Conversely, if there was any conflict, then the speculative work must be discarded. TM works well in cases where speculation pays off—i.e., where conflicts occur rarely in practice.

In the next section, we describe the general design choices that apply to different TMs. Then, in Section 11.3, we focus on HTM, examining Herlihy and Moss' HTM, and Moore et al.'s (2006) LogTM. Both these HTM systems modify the processor so that it tracks the speculative work that transactions do, and the designs rely on the processor to detect conflicts between transactions. In Section 11.4, we turn to STM, and show how data management and conflict detection can be built without requiring any changes to current hardware.

Programming directly over HTM or STM can be cumbersome; there is little portability between different implementations, and there can be a lot of boilerplate (e.g., the `do...while` loop and explicit `LT` and `ST` operations in Figure 11.3). Furthermore—and more seriously from a programming point of view—this kind of basic TM system provides atomic updates, but does not provide any kind of condition synchronization. This is a bit like a language providing locks but not providing condition variables: there is no way for a thread using TM to *block* while working on a shared data structure (e.g., for a `popLeft` operation on a deque to wait until an element is available to be removed).

Section 11.5 shows how these concerns can be tackled by providing `atomic` blocks in a programming language. We show how `atomic` blocks can be built over TM, and how the semantics of `atomic` blocks can be defined in a way that programs are portable from one implementation to another. Also, we show how `atomic` blocks can be coupled with constructs for condition synchronization.

Finally, in Section 11.6, we examine the performance of the Bartok-STM system, and the ways in which it is exercised by different transactional benchmarks.

11.2   Transactional Memory Taxonomy

A TM implementation must provide two main mechanisms: (1) a way to manage the speculative work that a transaction is doing as it runs, and (2) a way to detect conflicts between transactions, ensuring that only conflict-free transactions are committed to memory. At a high level, a distinction can be made between lazy version management and eager version management (Section 11.2.1), and between lazy conflict detection and eager conflict detection (Section 11.2.2).

11.2.1   Eager/Lazy Version Management

The eager/lazy version management distinction is about how a transaction manages its speculative updates. With eager version management, these updates are made directly to memory, and an undo log is maintained to record the values that are being overwritten. With lazy version management, the speculative work remains completely private to the transaction, and it is only written out to memory if the transaction completes without conflict. If conflicts are very rare, then there is an argument for eager version management: a transaction's updates are immediately available in memory.

11.2.2   Eager/Lazy Conflict Detection

A similar lazy/eager distinction is made about how conflicts are detected. With eager detection, a conflict is identified as soon as it occurs—say, if one transaction attempts to write to a location that a concurrent transaction has already written to.

Conversely, with lazy detection, the presence of a conflict might only be checked periodically—or it might only be detected at the end of the transaction when it attempts to commit.

The performance trade-offs between these approaches are more complicated than between lazy/eager version management. Under low contention, the performance of the conflict detection mechanism itself is more important than the question of whether it detects conflicts eagerly or lazy (in Section 11.4, we shall see examples of how this can influence the design of STM systems, where decisions have often been motivated by performance under low contention).

Under high contention, the way that conflicts are handled is of crucial importance, and most simple strategies will perform poorly for some workloads. Lazy conflict detection can cause threads to waste time, performing work that will eventually be aborted due to conflicts. However, eager conflict detection may perform no better: a transaction running in Thread 1 might detect a conflict with a transaction running in Thread 2 and, in response to the conflict, Thread 1 might either delay its transaction or abort it altogether. This kind of problem can be particularly pernicious if a conflict is signaled between two *running* transactions (rather than just between one running transaction and a concurrent transaction that has already committed). For instance, after forcing Thread 1's transaction to abort, Thread 2's transaction might itself be aborted because of a conflict with a third thread—or even with a new transaction in Thread 1. Without care, the threads can live-lock, with a set of transactions continually aborting one another.

TM implementations handle high contention in a number of ways. Some provide strong progress properties, for instance guaranteeing that the oldest running transaction can never be aborted, or guaranteeing that one transaction can only be aborted because of a conflict with a committed transaction, rather than because of a conflict with another transaction's speculative work (this avoids live-lock because one transaction can only be forced to abort because another transaction has succeeded). Other TMs use dynamic mechanisms to reduce contention, e.g., adding back-off delays, or reducing the number of threads in the system.

11.2.3 Semantics

Before introducing TM implementation techniques in Sections 11.3 and 11.4, we should consider exactly what semantics are provided to programs using TM. Current prototype implementations vary a great deal on all of these questions.

Access granularity: Does the granularity of transactional accesses correspond exactly to the size of the data in question, or can it spill over if the TM manages data at a coarser granularity (say, whole cache lines in HTM, or whole objects in STM)? This is important if adjacent data is being accessed non-transactionally—e.g., without precise access granularity, a TM using eager version management may roll back updates to adjacent non-transactional data.

Consistent view of memory: Does a running transaction see a consistent view of memory at all times during its execution? For example, suppose that we run the following transactions, with shared variables v1 and v2 both initially 0.

Transaction 1:   **Transaction 2:**

```
temp1 = LT(&v1);                          ST(&v1,1);
temp2 = LT(&v2);                          ST(&v2,1);
if (temp1 != temp2) while(1) {}           COMMIT();
COMMIT();
```

These two transactions conflict with one another, and so they cannot both be allowed to commit if they run in parallel. However, what if Transaction 2 runs in its entirety in between the two `LT` reads performed by Transaction 1? Some TMs guarantee that Transaction 1 will nevertheless see a consistent pair of values in `temp1` and `temp2` (either both 0, or both 1). Other TMs do not guarantee this and allow `temp1 != temp2`, so Transaction 1 may loop. This notion has been formalized as the idea of "opacity". When using TMs without opacity, programs must either be written defensively (so that they do not enter loops, or have other non-transactional side effects), or explicit validation must be used to check whether or not the reads have been consistent. Transactions that have experienced a conflict but not yet rolled back are sometimes called "zombies" or "orphans".

Strong atomicity: What happens if a mix of transactional and non-transactional accesses is made to the same location? Some TMs provide "strong atomicity", meaning that the non-transactional accesses will behave like short single-access transactions—e.g., a non-transactional write will signal a conflict with a concurrent transaction, and a non-transactional read will only see updates from committed transactions. Strong atomicity is typically provided by HTM implementations, but not by STM.

As we show in Section 11.5, `atomic` blocks can be defined in a way that allows portable software to be implemented over a wide range of different TMs, making these questions a concern for the language implementer (building `atomic` blocks over TM), rather than the end programmer (writing software using `atomic` blocks).

11.3   Hardware Transactional Memory

We sketched the use of TM in the example in Figure 11.3. In this section, we describe two different ways for that TM interface to be implemented in hardware. The first comes from Herlihy and Moss' paper that coined the phrase "transactional memory". It is an example of an HTM that uses lazy version management and eager conflict detection. The second design is LogTM, which provides an example of eager version management in hardware, also coupled with eager conflict detection.

11.3.1   Classical Cache-Based Bounded-Size HTM

Herlihy and Moss' insight was that conventional MESI data cache hardware already provides most of the facilities needed to make a series of memory accesses atomic. All that is required is to ensure that the locations being accessed remain in a private

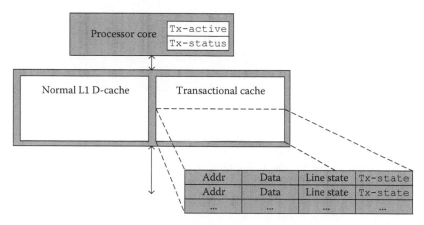

FIGURE 11.4: Herlihy and Moss' classical cache-based hardware transactional memory.

cache. The protocol used between caches remains largely unchanged, and committing or aborting a transaction can be done locally in the cache without extra communication with other processors or with main memory.

As Figure 11.3 shows, with Herlihy and Moss' HTM, distinct operations are used for transactional accesses: LT(p1) performs a transactional load from address p1, and ST(p2,x2) performs a transactional store of x2 to address p2. As an optimization, an additional LTX operation is available; this provides a hint that the transaction may subsequently update the location, and so the implementation should fetch the cache line in exclusive mode. A transaction is started implicitly by a LT/LTX/ST, and it runs until a COMMIT or an ABORT.

Within a transaction, any ordinary non-transactional memory accesses proceed as normal; the programmer is responsible for cleaning up the effects of any non-transactional operations that they do, and for making sure that non-transactional operations attempted by zombies are safe. The programmer is also responsible for invoking COMMIT at the end of the transaction (hence the while loop), and possibly adding an explicit delay to reduce contention.

Figure 11.4 sketches the implementation. The processor core is extended with two additional status flags. Tx-active indicates whether or not a transaction is in progress. If there is an active transaction, then Tx-status indicates whether or not it has experienced a conflict; although conflicts are detected eagerly by the hardware in Herlihy and Moss' design, they are not signaled to the thread until a VALIDATE or COMMIT.

The normal data cache is extended with a separate transactional cache that buffers the locations that have been accessed (read or written) by the current transaction. The transactional cache is a small, fully associative structure. This means that the maximum size of a transaction is only bounded by the size of the cache (rather than being dependent on whether or not the data being accessed maps to different cache sets).

Unlike the normal cache, the transactional cache can hold two copies of the same memory location. These represent the version that would be current if the running transaction commits, and the version that would be current if the running transaction aborts. The values held in any lower levels of the cache, and in main memory, all reflect the old value, and so this is a lazy versioning system. The two views of memory are distinguished by the Tx-state field in their transactional cache lines—there are four states: XCOMMIT (discard on commit), XABORT (discard on abort), NORMAL (data that was been accessed by a previously committed transaction, but is still in the cache), and EMPTY (unused slots in the transactional cache).

Conflicts can be detected by an extension of the conventional MESI cache protocol: a remote write to a line in shared mode triggers an abort, as does any remote access to a line in modified mode. In addition, a transaction is aborted if the transactional cache overflows, or if an interrupt is delivered.

Committing a transaction simply means changing the XCOMMIT entries to EMPTY, and changing XABORT entries to NORMAL. Correspondingly, aborting a transaction means changing the XCOMMIT entries to NORMAL, and changing XABORT entries to EMPTY. The use of XCOMMIT and XABORT entries can improve performance when there is temporal locality between the accesses of successive transactions— data being manipulated can remain in the cache, rather than needing to be flushed back to memory on commit and fetched again for the next transaction.

11.3.2   LogTM

LogTM is an example of an HTM system built on eager version management and eager conflict detection. Figure 11.5 illustrates the hardware structures that it uses. Unlike Herlihy and Moss' HTM, it allows speculative writes to propagate all the way out to main memory before a transaction commits: the cache holds the new values tentatively written by the transaction, while per-thread in-memory undo-logs record the values that are overwritten.

The processor core is extended to record the current base of the log and an offset within it. A single-entry micro-TLB provides virtual-to-physical translation for the current log pointer. This design allows the undo log entries to be cached in the normal way, and only evicted from the cache to main memory if necessary.

LogTM allows a transaction to be aborted immediately upon conflict, rather than continuing to run as a zombie. This is done by augmenting each thread's cache with additional metadata to detect conflicts between transactions, introducing R/W bits on each cache block. The R bit is set when a transaction reads from data in the block, and the W bit is set when a transaction writes to data in the block. A conflict is detected when the cache receives a request from another processor for access to a block in a mode that is incompatible with the current R/W bits—e.g., if a processor receives a request for write access to a line that it holds with the W bit set. When such a conflict is detected, either the requesting processor is stalled by sending a form of NACK message (eventually backing off to a software handler to avoid deadlock), or the transaction holding the data can be aborted before granting access to the requesting processor (in the case of a transaction updating the line, the update must be rolled back so that

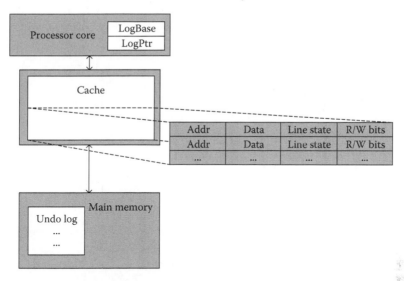

FIGURE 11.5: LogTM hardware overview. The data cache tracks read/write bits for each block, and buffers the new values being written by a transaction. The old values are held in an in-memory undo log so that they can be restored on abort.

the requester is provided with the correct contents). A transaction commits by flash clearing all the R/W bits and discarding the undo log.

In a directory-based implementation, LogTM can support transactions that overflow a processor's local cache by setting a special "Sticky-M" bit in the cache directory entry for a line that has been evicted by a transaction with its W bit set. The directory records which processor had accessed the line, and the protocol prevents conflicting access to the line until the transaction has committed or aborted.

LogTM-SE ("Signature Edition") provides decoupling between the structures used by the HTM and the L1 cache. Instead of detecting conflicts using per-block R/W bits, LogTM-SE uses read/write signatures that summarize the locations that have been read/written by each transaction. This avoids adding complexity to the cache, and it allows the cache to be shared between hardware threads without needing separate R/W bits for each thread. Signatures can be saved and restored by the operating system, allowing transactions to be preempted or rescheduled onto different cores. Furthermore, the operating system can maintain a summary signature representing the access of all currently descheduled threads, allowing conflict detection to be performed with them.

11.4 Software Transactional Memory

STM implementations can be built over the ordinary single-word atomic primitives provided by current hardware. Running a transaction in software is, of course,

```
Class Q {
  QElem leftSentinel;
  QElem rightSentinel;

  void pushLeft(int item) {
    QElem e = new QElem(item);
    do {
      TxStart();
      TxWrite(&e.right, TxRead(&this.leftSentinel.right));
      TxWrite(&e.left, this.leftSentinel);
      TxWrite(&TxRead(&this.leftSentinel.right).left, e);
      TxWrite(&this.leftSentinel.right, e);
    } while (!TxCommit());
  }
  ...
}
```

FIGURE 11.6:   Implementation of 'pushLeft' over an STM API.

not as streamlined as running it in hardware—the bookkeeping to track speculative updates and to detect conflicts must be handled explicitly by additional operations or function calls. However, by operating in software, STM provides flexibility for experimentation without hardware changes, and flexibility to support workloads that are not handled well by all HTM systems—for instance, very long running transactions.

Figure 11.6 shows the pushLeft operation written over an example STM API in which the accesses to shared memory are made by calls to TxWrite and TxRead functions that take the address of the word to access. The overall operation is wrapped in a loop that attempts to start a transaction (TxStart) and to commit it (TxCommit). As with the previous implementation, the memory allocation is done outside the loop so that it is not repeated.

Strong atomicity is typically not provided by STM systems—conflict detection only occurs between transactions, and unexpected results (and very implementation-dependent results) are seen when the same location is accessed concurrently in both modes. We return to this thorny question when looking at programming constructs in Section 11.5; for the moment, we write code so that each shared location is either always accessed transactionally, or always accessed non-transactionally.

11.4.1   Bartok-STM

Bartok-STM is an example of an STM that uses eager version management. It employs a hybrid form of conflict detection, with lazy conflict detection for reads, but eager conflict detection for writes. This follows from the use of eager version management: updates are made in-place, and so only one transaction can be granted write access to a location at any one time.

With Bartok-STM, there is no guarantee that a zombie transaction will have seen a consistent view of memory. The combination of this with eager version management means that the STM API must be used with care to ensure that a zombie does not write to locations that should not be accessed transactionally.

Bartok-STM maintains per-object metadata which is used for concurrency control between transactions. Each object's header holds a transactional metadata word (TMW) which combines a lock and a version number. The version number shows how many transactions have written to the object. The lock shows if a transaction has been granted write access to the object. A per-transaction descriptor records the current status of the transaction (`ACTIVE`, `COMMITTED`, `ABORTED`). The descriptor also holds an open-for-read log recording the objects that the transaction has read from, an open-for-update log recording the objects that the transaction has locked for updating, and an undo log of the updates that need to be rolled back if the transaction aborts.

The implementation of writes is relatively straightforward in Bartok-STM. The transaction must record the object and its current TMW value in the open-for-update log, and then attempt to lock the TMW by replacing it with a pointer to the thread's transaction descriptor. Once the transaction has locked the TMW, it has exclusive write access to the object.

The treatment of reads requires greater care. This is because Bartok-STM uses "invisible reads" in which the presence of a transaction reading from a location is only recorded in the transaction's descriptor—the reader is invisible to other threads. Invisible reading helps scalability by allowing the cache lines holding the TMW, and the lines holding the data being read, to all remain in shared mode in multiple processors' caches. The key to supporting invisible reads is to record information about the objects a transaction accesses, and then to check, when a transaction tries to commit, that there have been no conflicting updates to those objects. In Bartok-STM, this means that a read proceeds by (1) mapping the address being accessed to the TMW, and (2) recording the address of the TMW and its current value in the open-for-read log, before (3) performing the actual access itself. The decision to keep these steps simple, rather than actually checking for conflicts on every read, is an attempt to accelerate transactions in low-contention workloads.

When a transaction tries to commit, the entries in the open-for-read log must be checked for conflicts. Figure 11.7 illustrates the different cases. Figure 11.7a–c is the conflict-free case. In Figure 11.7a, the transaction read from an object and, at commit-time, the version number is unchanged. In the example, the version number was 100 in both cases. In Figure 11.7b, the transaction read from the object, and then the object was subsequently updated by the same transaction. Consequently, the transaction's open-for-update log and open-for-read logs record the same version number (100), and the TMW refers to the transaction's descriptor (showing that the object is locked for writing by the current transaction). In Figure 11.7c, the transaction opened the object for updating, and then subsequently opened it for reading: the TMW refers to the transaction descriptor, and the open-for-read log entry also refers to the transaction descriptor (showing that the object had already been written to by this transaction before it was first read).

The remaining cases indicate that a conflict has occurred. In Figure 11.7d, the object was read with version number 100 but, by commit-time, another transaction

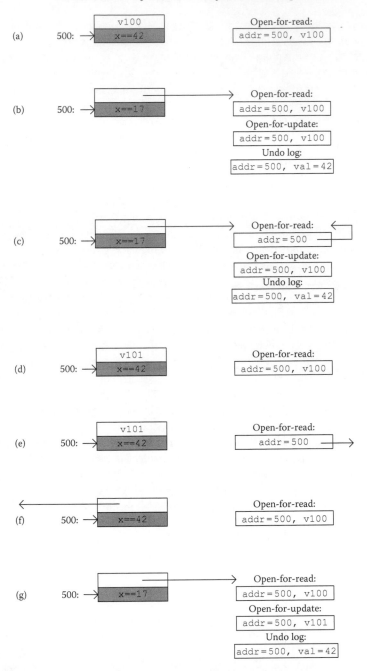

FIGURE 11.7: Read validation in Bartok-STM. In each case, the transaction has read from data at address 500. In cases (a)–(c), the transaction has not experienced a conflict. In cases (d)–(g), it has. The object itself is shown on the left, and the transaction's logs are shown on the right.

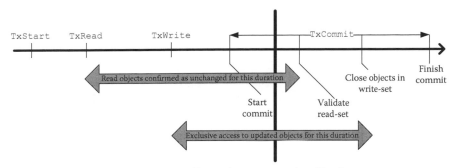

FIGURE 11.8: Linearization point of transactions using Bartok-STM.

has made an update and increased the version number to 101. In Figure 11.7e, the object was already open for update by a different transaction at the point when it was opened for reading: the open-for-read log entry recorded the other transaction's descriptor. In Figure 11.7f, the object was opened for update by another transaction concurrent with the current one: the object's TMW refers to the other transaction's descriptor. Finally, Figure 11.7g is the case where one transaction opened the object for reading, and later opened the same object for writing, but, in between these steps, a different transaction updated the object—the version number 100 in the open-for-read log does not match the version number 101 in the open-for-update log.

Given that the work of validation involves a series of memory accesses, how do we know that the transaction as a whole still has the appearance of taking place atomically? The correctness argument is based on identifying a point during the transaction's execution where (1) all of the locations read must have the values that were seen in them and (2) the thread running the transaction has exclusive access to all of the data that it has written. Such a "linearization point" is then an instant, during the transaction's execution, where it appears to occur atomically.

Figure 11.8 summarizes this argument: with Bartok-STM, the linearization point of a successful transaction is the start of the execution of the commit operation. At this point, we already know that the transaction has exclusive access to the data that it has written (it has acquired all of the locks needed, and not yet released any). Furthermore, for each location that it has read, it recorded the TMW's version number before the read occurred, and it will confirm that this version number is still up-to-date during the subsequent validation work.

11.4.2 TL2

The second STM system which we examine is TL2. This system takes a number of different design decisions to Bartok-STM: TL2 uses lazy version management, rather than maintaining an undo log, and TL2 uses lazy conflict detection for writes as well as reads. Furthermore, TL2 enables a transaction to see a consistent view of memory throughout its execution—because of this, and because lazy version management is used, the programmer does not have to worry about many of the problems of zombie transactions.

TL2 uses an elegant timestamp mechanism to check that a transaction's reads have seen a consistent view of memory—earlier STM algorithms could only provide this guarantee by sacrificing the use of invisible readers, or by adding per-access revalidation checks (causing a transaction reading from n distinct locations to need to perform $O(n^2)$ validation checks).

The main idea in TL2 is to maintain a global version number which is used to order transactions. When a transaction starts, it records the current value of the global version number into its transaction descriptor. This forms the transaction's read version (RV), and the TL2 algorithm ensures that a transaction does not see any values that are newer than this version; consequently, all of the values that a transaction sees form a consistent view of the heap at the point at which the transaction started.

The case of read-only transactions is straightforward: each read is performed by (1) checking that the object's TMW is unlocked and that the version number is no newer than RV, (2) reading the data, and (3) checking that the object's TMW is unchanged. This series of accesses takes a consistent snapshot of the TMW and the data itself. If either of the checks fails, then the transaction must be aborted. If a read-only transaction reaches the end of its execution, then it may commit without further validation checks; the time that the RV snapshot was taken forms the linearization point.

Read–write transactions are more complicated than read-only ones. The transaction descriptor holds two logs in addition to a record of its RV. The read log records the addresses of the TMWs that the transaction has read from (the actual values seen in the TMWs are not needed, but the read operations themselves must still check that the versions are no newer than RV). The write log records the speculative updates that the transaction wishes to make. Since lazy version management is used, reads must see values from the write set before reading from main memory; hashing into the write log can be used to accelerate this.

Figure 11.9 summarizes the way in which TL2 operates. As with Figure 11.8, it shows the points during a transaction's execution at which it first reads from an object, and when it first writes to an object, and then the individual steps involved in performing a commit operation.

When a transaction tries to commit, it locks the TMWs for the objects that it wishes to update. As with Bartok-STM's use of locks, this prevents concurrent access by other transactions trying to write to the same objects. If all of these locks are available, then the transaction obtains a write version (WV) by an atomic increment on the global version number. This step forms the transaction's linearization point.

Once the WV has been obtained, the transaction must check that the objects in its read set still hold version numbers that are no newer than RV. This confirms that there were no conflicts on objects in the read set from transactions that started after RV but before WV. Finally, if no conflicts are detected, the updates can be written back from the write log, and the TMWs can be unlocked by incrementing their version numbers to WV.

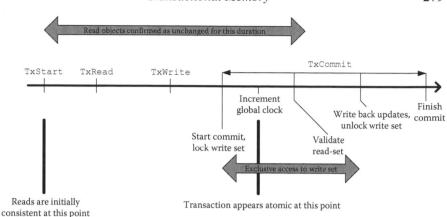

FIGURE 11.9: Linearization point of read-write transactions using TL2-STM.

11.5 Atomic Blocks

As Sections 11.3 and 11.4 have shown, TMs differ between hardware/software implementations (e.g., whether or not strong atomicity is provided), as well as between different HTM APIs (e.g., whether or not transactional operations must be explicitly marked), and between different STM APIs (e.g., whether or not the implementation provides a consistent view of memory at all times).

Adding atomic blocks to a language is one way to provide a higher-level programming abstraction that can be implemented in a portable manner across many of these TMs. Implementations of atomic blocks can be built over HTM, over STM, or by using static analyses to infer where an implementation should acquire and release internal locks. In much the same way, high-level languages with mutexes, condition variables, and volatile data provide an abstraction that is defined without needing to refer to the low-level details of a particular processor's virtual memory system, memory consistency model, and so on.

We return to the details of the semantics of atomic blocks in Section 11.5.1, but intuitively an atomic block can be thought of as running without any operations from other threads being interleaved in its execution. Figure 11.10 shows the implementation of pushLeft using an atomic block; the code inside the block is identical to the sequential version.

A naïve implementation of atomic blocks over STM can be built by (1) replacing each atomic block with a StartTx...CommitTx loop, iterating until the commit succeeds, (2) replacing each memory access within the transaction with a call on the appropriate ReadTx or WriteTx function, and (3), replacing each function call with a call to a cloned version of the function in which ReadTx and WriteTx are used. For example, in Figure 11.10, the constructor for QElem might need to be cloned in this way. Over HTM, with strong atomicity, the atomic block may be able

```
Class Q {
    QElem leftSentinel;
    QElem rightSentinel;

    void pushLeft(int item) {
        atomic {
            QElem e = new QElem(item);
            e.right = this.leftSentinel.right;
            e.left = this.leftSentinel;
            this.leftSentinel.right.left = e;
            this.leftSentinel.right = e;
        }
    }
    ...
}
```

FIGURE 11.10:   Implementation of 'pushLeft' using an atomic block.

to execute directly as a hardware transaction (with reads and writes replaced with transactional reads and writes, if the particular HTM interface requires this).

An implementation of atomic blocks must take care to handle object allocations. One approach is to view the actual allocation work to be part of a transaction implementing the atomic block, meaning that the allocation is undone if the transaction is rolled back. However, this can introduce false contention (e.g., if allocations of large objects are done from a common pool of memory, rather than per-thread pools). An alternative approach is to integrate the language's memory allocator with the TM implementation, tracking tentative allocations in a separate log, and discarding them if a transaction rolls back.

In a system using a garbage collector, tentative allocations can simply be discarded if a transaction rolls back: the GC will reclaim the memory as part of its normal work. However, the GC implementation must be integrated with the TM—e.g., scanning the TM's logs during collection, or aborting transactions that are running when the collector starts.

11.5.1   Semantics of Atomic Blocks

At first glance, the semantics of atomic blocks looks straightforward. However, intuition can be deceptive and so a more precise definition is needed. Figure 11.11 illustrates a "privatization" idiom which helps illustrate some of the problems. In this example, the programmer may intend that o2.val is accessed inside atomic blocks when o1.isShared is true, but that o2.val is private to Thread 2 when o1.isShared is false.

Early implementations of atomic blocks defined their semantics in terms of transactions. This style of definition limits portability—although the syntax looks like it abstracts the details of the STM, the semantics does not. For example, the idiom in Figure 11.11 works over most HTM systems, but not over most early STMs—e.g.,

```
// Thread 1
atomic {
  if (o1.isShared) {
    o2.val++;
  }
}

// Thread 2
atomic {
    o1.isShared = false;
}
o2.val = 100;
```

FIGURE 11.11: A 'privatization' idiom in which Thread 1 accesses o2.val from an atomic block, and Thread 2 accesses o2.val directly.

with Bartok-STM, Thread 1 could update o2.val from a zombie transaction, even after Thread 2's transaction has finished. Conversely, idioms that rely on some particular quirks of a given STM may not work with HTM.

Arguably, a cleaner approach is to define the semantics of atomic blocks independently of the notion of transactions. For example, the idea of "single lock atomicity" (SLA) requires that the behavior of atomic blocks be the same as that of critical sections that acquire and release a single process-wide lock. With this model, the privatization idiom is guaranteed to work (because it would work with a single lock). Many of the examples that only "work" with STM will involve a data race when implemented with a single global lock, and so their behavior would be undefined in many programming languages.

To support atomic blocks with SLA, it is necessary that granularity problems do not occur, that the effects of zombie transactions are not visible to non-transacted code, and that ordering between transactions (say, Tx1 is serialized before Tx2) ensures ordering of the surrounding code (code that ran before Tx1 in one thread must run before code after Tx2 in another thread).

An alternative approach to SLA is to define the semantics of atomic blocks independently from TM and from existing constructs. This may require additional work when designing a language (or when learning it), but is more readily extendable to include additional constructs such as operations for condition synchronization. Following this approach, typical definitions require a "strong semantics" in which an atomic block in one thread appears to execute without any interleaving of work from other threads—i.e., no other work at all, neither other atomic blocks, nor normal code outside atomic blocks. The "appears to" is important, of course, because the definition is not saying that atomic blocks actually run serially, merely that the program will behave as if they do so.

If it can be implemented, then such strong semantics would mean that the programmer does not need consider the details of particular TM implementations—or indeed the question of whether atomic blocks are implemented optimistically using TM, or via some kind of automated lock inference.

The basic implementation we have sketched clearly does not provide strong semantics to all programs, because it does not implement examples like the privatization idiom when built over the STM systems from Section 11.4. Furthermore, it is not even the case that an implementation built over HTM with strong atomicity would run all programs with strong semantics because of the effect of program transformations during compilation, or execution on a processor with a relaxed memory model. This dilemma can be reconciled by saying that only "correctly synchronized" programs need to execute with strong semantics; this is much the same as when programming with locks, where only data-race-free programs are typically required to execute with sequential consistency.

With `atomic` blocks, there is a trade-off between different notions of "correct" synchronization, and the flexibility provided to use a wide range of TM implementations. At one extreme, STM-Haskell enforces a form of static separation in which transactional data and non-transactional data are kept completely distinct; the type system checks this statically, and all well-typed STM-Haskell programs are correctly synchronized. This provides a lot of flexibility to the language implementer, but the simple type system can make code reuse difficult, and require explicit marshalling between transactional and normal data structures. For example, the privatization idiom is not well-typed under static separation.

An alternative notion of correct synchronization is to support transactional data-race-free (TDRF) programs. Informally, a program is TDRF if, under the strong semantics, there are no ordinary data races, and there are no conflicts between accesses from normal code and code inside `atomic` blocks. This is modeled on the conventional definition of data-race freedom from programming language memory models. The privatization idiom from Figure 11.11 is TDRF.

Given the need to consider notions of correct synchronization in defining the semantics of `atomic` blocks, is it actually fair to say that they provide an easier programming model than using explicit locks? That is a question that must ultimately be tested experimentally, but intuitively, even if the programming model is a form of SLA, programming with a single lock is simpler than programming with a set of locks; the question of exactly which lock to hold becomes the question of whether or not to hold the single lock.

11.5.2   Optimizing Atomic Blocks

A basic implementation of `atomic` blocks can provide dreadful performance if TM is used for all of the memory accesses that are made—even with HTM there can be a cost of additional work when writing to an undo log, or additional pressure on caches to hold multiple versions of the same data.

Several significant improvements are possible. First, variables on the stack are usually guaranteed to be thread-local: an implementation may still need to log their values for rollback, but conflict detection is not needed. It is easy to treat local variables as a special case in C# and Java because special bytecodes are used for most accesses to the stack. Second, static analyses can identify some cases where objects are local to an `atomic` block (e.g., the new QElem object in Figure 11.10). Finally, many

```
Class Q {
  QElem leftSentinel;
  QElem rightSentinel;

  void pushLeft(int item) {
    do {
      TxStart();
      QElem e = new QElem(item);
      TxOpenForRead(this);
      TxOpenForWrite(this.leftSentinel);
      e.right = this.leftSentinel.right;
      e.left = this.leftSentinel;
      TxOpenForWrite(this.leftSentinel.right);
      TxLogForUndo(&this.leftSentinel.right.left);
      this.leftSentinel.right.left = e;
      TxLogForUndo(&this.leftSentinel.right);
      this.leftSentinel.right = e;
    } while (!TxCommit());
  }
  ...
}
```

FIGURE 11.12: Optimizing placement of STM operations using an STM with object-based conflict detection, and eager version management.

workloads exhibit a strong degree of locality within a transaction and so redundant TM operations can be caused by repeated loads/stores to the same data.

These observations can be exploited by decomposing the interface to the STM, so that the bookkeeping operations are separated from the actual data accesses. As before, TxStart and TxCommit bracket the transaction. However, the reads and writes are preceded by TxOpenForRead (for reads) and TxOpenForWrite and TxLogForUndo (for writes). These operations can then be moved independently of the data accesses based on knowledge of the particular STM, e.g., if concurrency control is done on a per-object basis, then one TxOpenForRead will provide access to multiple fields.

Figure 11.12 illustrates the use of these operations when building over an STM with eager version management. No concurrency control is needed on the local variable "e", or on the fields of the QElem object that "e" refers to. The object that "this" refers to must be opened for reading, and the objects that "this.leftSentinel" and "this.leftSentinel.right" refer to must be opened for writing (so that the new QElem can be spliced between them).

11.5.3 Composable Blocking

We have shown how atomic blocks can be used to update data structures in shared memory. What about cases where a programmer wishes one thread to wait until the contents of memory change in some way—perhaps for a version of `popLeft` that blocks if it finds that the deque is empty?

STM-Haskell provides a pair of operations known as `retry` and `orelse` to tackling this kind of problem. Calling `retry` indicates that the current `atomic` block is not yet ready for run, e.g., in C#-like syntax:

```
int popLeft() {
 atomic {
  if (this.leftSentinel.right == this.rightSentinel) retry;
  ...
 }}
```

Unlike when programming with locks and condition variables, it is not necessary to identify what condition `popLeft` is dependent on, or what condition will cause it to succeed in the future. This avoids lost-wake-up problems where a variable is updated, but threads waiting on conditions associated with the variable are not signaled. With STM-Haskell, the thread running the `atomic` block waits until an update is committed to any of the locations that the `atomic` block has read.

This form of blocking is composable in the sense that an `atomic` block may call a series of operations that might wait internally, and the `atomic` block as a whole will only execute when all of these conditions will succeed. For example, to take two items:

```
atomic {
 x1 = q.popLeft();
 x2 = q.popLeft();
}
```

The combined `atomic` block can only complete when both `popLeft` calls return items, and atomicity requires that the two items be consecutive. Of course, the same structure could be used for other operations—taking more than two items, or taking elements that meet some other requirement (say, taking a different number of items dependent on the value of x1). All these alternatives can be built without changing the underlying deque.

Having said that, `retry` must be used with care—the programmer must ensure that it will actually be possible for the operations to execute together; it is no good enclosing two `popLeft` calls in an `atomic` block if the underlying buffer is built over a single storage cell. Similarly, sets of possibly blocking operations that involve communication with other threads cannot generally be performed atomically—e.g., updating one shared buffer with a request to a server thread, and then waiting for a response to arrive. The server's work must happen in between the request and the response, so the request–response pair cannot happen atomically.

An `orelse` construct provides a way to try a piece of code and to catch any attempt is makes to block, e.g., to turn `popLeft` into an operation that returns a failure code instead of waiting:

```
int popLeftNoWait() {
  atomic {
    return popLeft();
  } orelse {
    return -1;
  }}
```

The semantics of `orelse` is that either (1) the first branch executes as normal, or (2) the first branch reaches `retry`, in which case the second branch executes in its place (i.e., any putative updates from the first branch are discarded). If both branches call `retry`, then the `retry` propagates, as with an exception, out to an enclosing `orelse`. Alternative constructs could be defined which provide a non-deterministic choice between their branches, rather than the left-biased form provided by `orelse`—however, note that the left bias is essential for the usage in `popLeftNoWait`.

11.6 Performance

The design and implementation of TM systems remains an active research topic, and numerous prototype systems have been described in the literature, or are available for experimental use.

As with any parallel algorithm, several different metrics are interesting when evaluating the performance of STM systems. If a system is to be useful in practice, then the *sequential overhead* of using transactions must not be so great that a programmer is better off sticking with single-threaded code. For instance, if transactional code runs 16× slower than single-threaded code, then a program would need to be able to scale perfectly to 16 cores just to recoup this loss; contention in the memory hierarchy, conflicts between transactions, uneven work distribution between threads, and synchronization elsewhere in the language runtime system would make this kind of scaling difficult to achieve. Sequential overhead is largely affected by the costs introduced by the TM system; e.g., additional bookkeeping in an STM system, or additional pressure on cache space in an HTM system. The sequential overhead of an STM is highly dependent on the performance of the baseline language implementation, and on the engineering and tuning that has been employed in building the STM itself; a poor language implementation can mask the high costs of a given STM system.

In addition to sequential overhead, a programmer should be interested in the way in which a particular TM implementation affects the scalability of their programs. Scalability is a function of both the program's workload (e.g., if the transactions involve conflicts), and the internals of the TM system (e.g., whether or not the implementation introduces synchronization between unrelated transactions).

In this section, we briefly examine the performance achieved by the Bartok-STM system. The implementation operates as an ahead-of-time compiler from C# to native

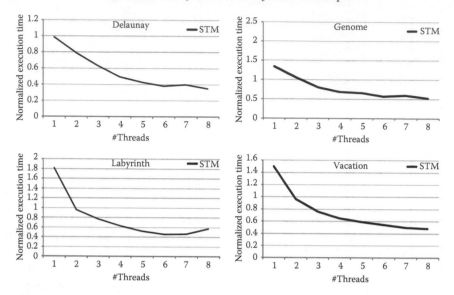

FIGURE 11.13: Performance of the Bartok-STM implementation.

x86 code, employing standard program transformations to optimize the C# code, and using the techniques from Section 11.5.2 to optimize the placement of STM operations within transactions. Other STM systems can provide lower sequential overhead (e.g., NOrec), or better scalability (e.g., SkySTM), but we stick with Bartok-STM to complete the description of a full system.

Our results look at a set of four benchmarks. Three of these were derived from the STAMP 0.9.9 benchmark suite (Genome, Labyrinth, Vacation). We translated the original C versions of these programs into C#, making each C `struct` into a C# `class`. The fourth benchmark is a Delaunay triangulation algorithm implemented following the description by Scott et al. (2007). In all four benchmarks, we added `atomic` blocks to the source code, and the compiler automatically added calls to the STM library for code within the atomic blocks, before automatically optimizing the placement of the STM operations.

Figure 11.13 shows the results on a machine with two quad-core Intel Xeon 5300-series processors. All results show execution time for a fixed total amount of work, normalized against execution of a sequential program running on a single core. This means that the 1-thread STM results show the sequential overhead that is incurred, and the point at which the STM curves cross 1.0 shows the number of threads that are needed in order to recoup this overhead. For all of these workloads, the STM-based implementation out-performs the sequential version when two threads are used.

These workloads vary a great deal. Delaunay uses short transactions when stitching together independently triangulated parts of the space: most of the program's execution is during non-transactional phases in which threads work independently. As Scott et al. (2007) observed, one way that TM can affect performance for this workload is if it imposes any overheads on non-transactional execution—e.g., by adding an additional level of indirection to each object access. Bartok-STM behaves well from this point of view: normal memory accesses can be used during the non-transactional phases of execution.

The Genome workload spends around 60% of its time in transactions. These are modestly sized operations, accessing 17 memory locations on average. The larger time spent inside transactions leads to a higher sequential overhead.

Labyrinth is unique amongst these programs in that it spends almost all of its time inside transactions, and it executes a small number of extremely large transactions which access tens of thousands of locations. The benchmark performs routing within a 3D maze, with each transaction attempting to route between a given source–destination pair in the maze. Transactions detect conflicts when the routes proposed by different threads collide. The size of the transactions, and the conflict rate, depends on the initial dimensions of the maze and the number of source–destination pairs to be routed. The benchmark is also notable in demonstrating the use of an "early release" programming abstraction in which parts of a transaction's data set are discarded during its execution. This is done to provide an approximate snapshot of the maze at the start of the transaction's work, allowing the transaction to read from all of the locations in the maze without incurring conflicts on locations that it does not subsequently use in its chosen route. Without this optimization, essentially every pair of transactions would conflict.

The final benchmark, Vacation, maintains a simple in-memory database represented by a series of trees. These are accessed to model booking and querying travel arrangements. The transactions access an average of 57 locations, and around 68% of the program's single-threaded execution time is spent in transactions. This benchmark illustrates a workload which would be difficult to manage using fine-grained locking: even if an individual tree was built in a scalable way, it would be necessary to add an additional layer of locking to provide atomicity between related updates to different trees.

11.7 Where Next with TM?

In this chapter, we have introduced three main areas of research: HTM, STM, and `atomic` blocks. There are evidently common aspects to all three, but it is worthwhile highlighting the differences between them and, in particular, the differences in the arguments that might be used when deploying them in a mainstream language or in an actual piece of hardware.

There is a clear argument that even a modestly sized HTM—2, 3, or 4 locations— provides a simplification to the implementation of non-blocking shared-memory data structures, and a performance benefit over known software implementations of the data structures. It allows a small set of locations to be accessed without needing to handle ABA problems, and without the usually attendant costs of bookkeeping or dynamic memory management. With this approach, low-level libraries could encapsulate the use of transactions for a data structure, using HTM when it is available, and otherwise falling back to a specialized software implementation of the data structure. This is similar to the use of other architecture-specific techniques, say SSE4 instructions.

The argument for `atomic` blocks in a general-purpose language is different. The aim there is to provide a higher-level abstraction for building concurrent data structures; in any particular case it is very likely that a specialized design can perform better—but potentially losing the composability of different data structures built over transactions, or the ability to control blocking via constructs like `orelse`. Of course, for `atomic` blocks to be useful in practice, the performance must be sufficiently good that the cost of using a general-purpose technique rather than a specialized one is acceptable. Currently, the state of the art in STM is perhaps akin to garbage collection in the early 1990s; a number of broad design choices are known (eager version management vs. lazy version management on the one hand, copying vs. mark-sweep vs. reference counting on the other), but there is still work to be done in understanding how to build a high-performance implementation and, in particular, one that performs predictably across workloads.

An alternative, more incremental, approach is to use TM within the implementation of existing language constructs. Forms of speculative lock elision have been studied in hardware and in software; the idea is to allow multiple threads to execute critical sections speculatively, allowing them to run in parallel so long as they access disjoint sets of memory locations. TM can form the basis of an implementation, and the question of whether to use TM, or whether to use an actual lock, can be based on the performance characteristics of the particular TM implementations that are available.

11.8 Chapter Notes

The work described in this chapter is based on a wide range of research papers. The brief notes here are meant as pointers to some of these papers, rather than a full survey. Harris, Larus, and Rajwar's (2010) book provides a more detailed discussion of work in this area.

Herlihy and Moss' (1993) coined the term transactional memory and developed the classical implementation of bounded transactions that we describe in Section 11.3. Moore et al. (2006) designed the LogTM algorithm which pioneered the use of eager updates in HTM.

Shavit and Touitou (1995) described the first software implementation of transactional memory. It provides a static interface (i.e., one in which a set of memory locations and proposed updates are presented in an array passed to a single operation). Herlihy et al.'s (2003) DSTM is an early practically focused non-blocking object-based design. The Rochester Synchronization Group's RSTM is a mature research prototype STM library for C++.

Bartok-STM and McRT-STM introduced the combination of eager version management and lazy conflict detection on reads. These systems have been used to explore the application of static analyses to optimize the placement of calls onto a TM library. Dice et al (2006). designed the original TL2 algorithm.

Two examples of recent STM systems which illustrate different design choices are NOrec (Dalessandro et al. 2010) and SkySTM (Lev et al. 2009). The NOrec STM system provides an example of design choices taken to reduce the sequential overheads incurred by an STM. It avoids the need to maintain any per-object or per-word metadata for conflict detection. Instead, transactions maintain a value-based log of their tentative reads and writes, and use commit-time synchronization to check whether or not these values are still up-to-date. The SkySTM system demonstrates a series of design choices to provide scalability to multiprocessor CMP systems comprising 256 hardware threads. It aims to avoid synchronization on centralized metadata, and employs specialized *scalable nonzero indicator* (SNZI) structures to maintain distributed implementations of parts of the STM system's metadata.

Harris et al. (2005, 2006) introduced the `retry` and `orelse` constructs in GHC-Haskell and provided an operational semantics for a core of the language. Moore et al. (2006) and Abadi et al. (2008) provided formal semantics for languages including atomic actions. They showed that the static separation programming discipline allowed flexibility in TM implementation, and they showed that a simple type system could be used to ensure that a program obeys static separation.

Shpeisman et al. (2007) provided a taxonomy of problems that occur when using STM implementations that do not provide strong atomicity. Menon et al. (2008) studied the implementation consequences of extending such an implementation to support SLA.

References

Abadi, M., A. Birrell, T. Harris, and M. Isard, Semantics of transactional memory and automatic mutual exclusion, *Proceedings of the 35th Annual Symposium on Principles of Programming Languages, POPL 2008*, San Francisco, CA.

Adl-Tabatabai, A.-R., B.T. Lewis, V. Menon, B.R. Murphy, B. Saha, and T. Shpeisman, Compiler and runtime support for efficient software transactional memory, *Proceedings of the 2006 Conference on Programming Language Design and Implementation, PLDI 2006*, Ottawa, Ontario, Canada.

Dalessandro, L., M.F. Spear, and M.L. Scott, NOrec: Streamlining STM by abolishing ownership records, *Proceedings of the 15th Symposium on Principles and Practice of Parallel Programming, PPoPP 2010*, Bangalore, India.

Dice, D., O. Shalev, and N. Shavit, Transactional locking II, *Proceedings of the 20th International Symposium on Distributed Computing, DISC 2006*, Stockholm, Sweden.

Harris, T., J. Larus, and R. Rajwar, *Transactional Memory*, 2nd edn., Morgan & Claypool Publishers, San Rafael, CA, 2010.

Harris, T., S. Marlow, S. Peyton Jones, and M. Herlihy, Composable memory transactions, *Proceedings of the 10th Symposium on Principles and Practice of Parallel Programming, PPoPP 2005*, Chicago, IL.

Harris, T., M. Plesko, A. Shinnar, and D. Tarditi, Optimizing memory transactions, *Proceedings of the 2006 Conference on Programming Language Design and Implementation, PLDI 2006*, Ottawa, Ontario, Canada.

Herlihy, M. and J.E.B. Moss, Transactional memory: Architectural support for lock-free data structures, *Proceedings of the 20th International Symposium on Computer Architecture, ISCA 1993*, San Diego, CA.

Herlihy, M., V. Luchangco, M. Moir, and W.N. Scherer III, Software transactional memory for dynamic-sized data structures, *Proceedings of the 22nd Annual Symposium on Principles of Distributed Computing, PODC 2003*, Boston, MA.

Lev, Y., V. Luchangco, V. Marathe, M. Moir, D. Nussbaum, and M. Olszewski, Anatomy of a scalable software transactional memory, *TRANSACT 2009*, Raleigh, NC.

Menon, V., S. Balensiefer, T. Shpeisman, A.-R. Adl-Tabatabai, R.L. Hudson, B. Saha, and A. Welc, Practical weak-atomicity semantics for Java STM, *Proceedings of the 20th Symposium on Parallelism in Algorithms and Architectures, SPAA 2008*, Munich, Germany.

Moore, K.E., J. Bobba, M.J. Moravan, M.D. Hill, and D.A. Wood, LogTM: Log-based transactional memory, *Proceedings of the 12th International Symposium on High-Performance Computer Architecture, HPCA 2006*, Austin, TX.

Moore, K.F. and D. Grossman, High-level small-step operational semantics for transactions, *Proceedings of the 35th Annual Symposium on Principles of Programming Languages, POPL 2008*, San Francisco, CA.

Scott, M.L., M.F. Spear, L.Dalessandro, and V.J. Marathe, Delaunay triangulation with transactions and barriers, *Proceedings of the 2007 IEEE International Symposium on Workload Characterization*, Boston, MA.

Shavit, N. and D. Touitou, Software transactional memory, *Proceedings of the 14th Annual Symposium on Principles of Distributed Computing, PODC 1995*, Ottawa, Ontario, Canada.

Shpeisman, T., V. Menon, A.-R. Adl-Tabatabai, S. Balensiefer, D. Grossman, R.L. Hudson, K.F. Moore, and B. Saha, Enforcing isolation and ordering in STM, *Proceedings of the 2007 Conference on Programming Language Design and Implementation, PLDI 2007*, San Diego, CA.

Chapter 12

Emerging Applications

Pradeep Dubey

Contents

12.1   Introduction

A wave of digitization is all around us. Digital data continues to grow in leaps and bounds in its various forms, such as unstructured text on the Web, high-definition images, increasing digital content in health clinics, streams of network access logs or ecommerce transactions, surveillance camera video streams, as well as massive virtual reality datasets and complex models capable of interactive and real-time rendering, approaching photo-realism and real-world animation. While none of us perhaps has the crystal ball to predict the future *killer app*, it is our belief that the next round of killer apps will be about addressing the *data explosion* problem for end users, a problem of growing concern and importance for both corporate and home users. The aim of this chapter is to understand the nature of such applications and propose a

Recognition   Mining   Synthesis

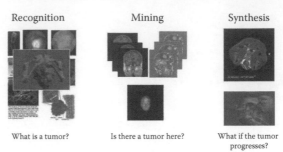

What is a tumor?   Is there a tumor here?   What if the tumor
                                                    progresses?

FIGURE 12.1:   Illustrating RMS components.

structured approach to applications analysis that can offer a deeper understanding of
the underlying software and hardware implications.

We begin by taking a deeper look at the computational needs of emerging applications. Most applications can be thought of in terms of a set of objects or events.
Consider, for example, the following three applications: finding a flower in an image
database, detecting an intrusion in a network, or synthesizing a castle in a virtual
world. *Flower*, *intrusion*, and *castle* are respective examples of target end-user object
or event in these illustrative cases. A typical implementation alternative for such
applications begins with an assumed computational model for the target object or
event. For example, a flower can be thought of as a combination of properly textured
polygons, whereas an intrusion can be modeled as a class of anomalous network
access patterns. We refer to this phase as the modeling or *recognition* (R) phase.
Given a model of a *flower*, *intrusion*, or *castle* makes it possible for the computer
to mine instances similar to a query instance through multidimensional indexing of
a resident or streaming dataset. This is referred to as the *mining* (M) phase. Finally,
a good computational model of a *flower*, *intrusion*, or *castle* is also a key to being
able to synthesize a virtual near-neighbor instance of the query object or simulate
a future progression of the event query. This computational phase is referred to as
the *synthesis* (S) phase. Figure 12.1 illustrates this recognition, mining, and synthesis (RMS) taxonomy through an example application. For a more detailed discussion
of such applications, the reader is referred to [1]. This unified taxonomy of RMS
helps us understand the computational needs of emerging compute-intensive applications in a structured manner. Section 12.2 offers a more detailed look at RMS components. It also discusses the impact of the Internet and massive Web datasets for
RMS computing. This is followed by Section 12.3 on hardware and software system
implications.

12.2   RMS Taxonomy

At a high level, recognition is often a type of machine learning. Computers need
to be able to examine data and images, and construct mathematical models based on
what they "see." Say, for example, the Black–Scholes model of an equity pricing, or

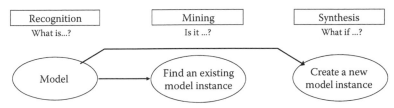

FIGURE 12.2: RMS taxonomy.

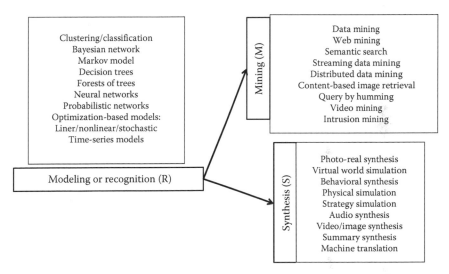

FIGURE 12.3: Decomposing emerging applications.

a data-acquisition-based model of the object or event, such as a collection of images believed to represent a certain flower type. As shown in Figure 12.2, recognition is the "what is." It is about identifying a set of data that constitutes a statistical model or constructing an analytical model. Some of the commonly used mathematical models are enumerated in Figure 12.3. Recognition needs to be a continuous process because data is always being created, always coming in. What's more, through constant model building, computers will get better at enriching a model with further data and eliminating the data that is not necessary.

Once the computer has recognized the "what is" and turned it into a model, the computer must be able search for instances of the model, or "is it" tasks. We label these as "mining" tasks. Mining refers to searching a dataset, such as the Web, to find instances of a given query model. The better computers are able to build models (recognition), the better computers should be at finding instances that fit these models and our needs. The primary computational challenge for this phase of computing is *multidimensional data indexing*. Given a large set of attributes of a multimodal query object (such as colors, textures, shapes, etc.), real-time indexing challenge is nontrivial. Some of the emerging mining applications are enumerated in Figure 12.3.

Synthesis is discovering "what if" cases of a model. If an instance of the model does not exist, a computer should be able to create a potential instance of that model in an imaginary world. In other words, synthesis is the ability to create an instance of a model where one does not exist, or simulate (predict) a future instance of an event sequence. Graphics is one of the most popular synthesis applications aimed at generating photo-real instances of real-world objects. Figure 12.3 lists a variety of various synthesis application classes.

In Section 12.2.1, we introduce the motivation for viewing the three RMS components as a loop, and introduce two most popular instances of RMS loop from the domains of visual computing and analytics usages. It should be clear from the introduction earlier in this section that *models* (of objects or events) play a critical role in RMS taxonomy. Section 12.2.2 takes a deeper look at a class of models of growing importance: data-driven or statistical models. Section 12.2.3 introduces a third instance of RMS loop, referred to as *nested RMS*, that models a growing class of Web-based, real-time, service-oriented, computing usages. Finally, Section 12.2.4 proposes a structured decomposition of RMS applications in an effort to offer a unified *worldview* of seemingly diverse emerging applications. We then move on to discussing the common system implications of RMS applications in Section 12.3. As we progress through the chapter, it should become increasingly clear that while many of the RMS applications are similar to the traditional HPC (high performance computing) application, they are significantly different in many ways. The primary difference stems from the fact that these are driven by mass usages, such as entertainment and enterprise productivity, as opposed to scientific discovery. This further has significant implications on underlying algorithm and fidelity requirements. Also, the emerging Web-based compute infrastructure with massive data, ubiquitous connectivity, and teraflops-class general-purpose multicore/manycore computing processors is quite different from the traditional supercomputer-based HPC infrastructure.

12.2.1   Interactive RMS (iRMS)

An end user judges the goodness of a computer model of an object or event only indirectly, through the quality of mining and synthesis outputs (Figure 12.4). For example, a search engine that returns excellent matches to a given *flower* input must have a very good computational model of flower. Similarly, an intrusion detection system which successfully detects a *security breach* must have a good predictive model of ensuing intrusions. More importantly, there is often a feedback loop (Figure 12.5) from the mining and synthesis output back to the modeling phase to continuously refine the model. Computational power that can enable real-time or interactive iteration of a model refinement task has the potential to create new end-user applications. For example, interactive volume rendering and surface extraction can enable medical imaging to move from mostly diagnostic imaging today (such as CT and PetScan) to interventional imaging (such as computer-assisted surgery).

For better understanding of various emerging applications, let us look at the iRMS loop in two different use contexts: visual computing and real-time analytics. Visual

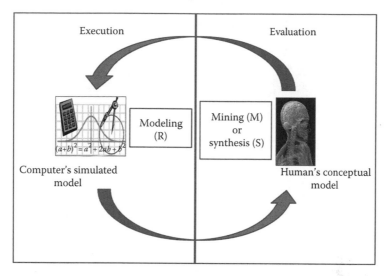

FIGURE 12.4: Mining or synthesis quality determines the goodness of a model for an end user.

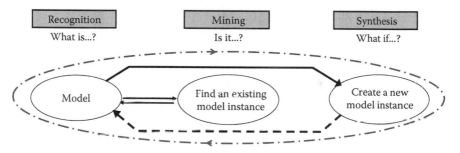

FIGURE 12.5: Most RMS apps are about enabling interactive (real-time) RMS loop (iRMS).

computing is an interdisciplinary combination of, primarily, graphics rendering, physical simulation, and computer vision. iRMS loop in this case is derived from the real-time integration of graphics and computer vision pipelines. Graphics pipeline transforms a polygonal model to a time sequence of images, whereas, vision pipeline does the inverse operation of transforming an image sequence into a model. This is illustrated in Figure 12.6. Analytics loop, illustrated in Figure 12.7, crawls a large amount of streaming or archived data to continuously derive richer ontologies capable of real-time predicting or simulating likely "what-if" scenarios, and thus forms the necessary framework of real-time analytics, for example, program trading applications in financial analytics.

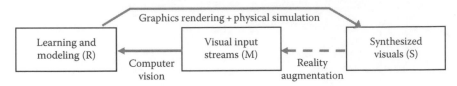

FIGURE 12.6:   Visual computing loop.

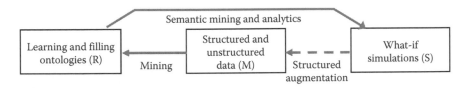

FIGURE 12.7:   Analytics loop.

12.2.2   Growing Significance of Data-Driven Models

Figures 12.8 and 12.9 show unrolled versions of visual computing and analytics loops discussed in the preceding text, respectively, to highlight the significant constituent compute functions. Common to these loops are two classes of models: (1) models that have a well-defined analytical or procedural formation, such as the Black–Scholes model of an equity pricing, or the Navier–Stokes formulation of fluid flow, and (2) data-acquisition-based statistical model, such as a collaborative filter model. Visual computing has predominance of procedural and analytical models as the physical phenomenon behind photo or physical realism are very well understood. The same cannot be said for nonvisual computing applications such as Web mining or financial analytics. These applications are dominated by data-driven statistical models. The quality of such data-based models is critically dependent on having large enough data input. Recent rapid growth of digital data and its easy accessibility has been a boon for this class of models. Real-time availability of acquired data can be further used to incrementally train and improve model accuracy. Model training in the past, such as that in the case of speech recognition, has often proven very difficult due to the limited amount of training data on a client machine. Furthermore, limited to an individual user of a client machine, the duration of a training loop to achieve an acceptable level of accuracy has been unacceptable as well. On the other hand, online databases, such as Flicker and YouTube, contain a massive amount of data, many times more than any client machine, and any Web-based system has the potential of training itself from a vast number of online users. Consequently, online systems with large datasets, such as Google search engine, have proven to be very successful. Real-time connectivity implies continuous improvement in model accuracy with new data and training inputs. Even for traditional databases, connectivity, coupled with growth in memory capacity and compute capability, opens the possibility of bringing together the current decoupled worlds of transaction and analytics, say, through in-memory real-time databases.

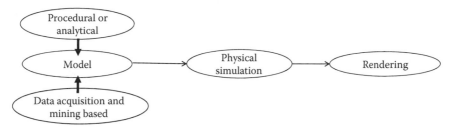

FIGURE 12.8: Visual computing loop (graphics and vision).

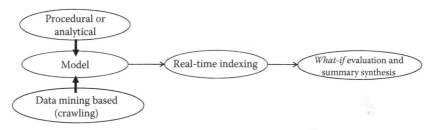

FIGURE 12.9: Search and analytics loop.

12.2.2.1 Massive Data Computing: An Algorithmic Opportunity

Availability of massive data has also opened up new superior algorithmic alternatives to many traditional problems. Two such illustrative applications are described in the following text:

- *Language-independent statistical machine translation:* Whereas the principles of statistical machine translation have been known for many years, most available translation systems make very little use of it, and instead focus on building a language grammar-specific translation. In somewhat of a surprise to this research community, a recent Google's English to Chinese and Arabic translator [2] built on the statistical approach outperformed all its language-specific counterparts. Key to the recent success of this approach can be attributed to the *massive* training data of human-translated books and documents. Aided by this data, it was feasible to build a high enough accuracy translation without any knowledge of the specific languages involved.

- *Scene completion:* A basic photo-editing operation involves removing an unwanted part of an image, and then filling the hole through various interpolation techniques involving data from nearby pixels and segments, aimed at creating a natural look. For example, removing an unwanted car parked in front of a building would involve filling the hole using texture and patches of the now exposed building. Hayes and Efros [3] present an alternate algorithmic approach to this classic problem based on a *massive* database of online

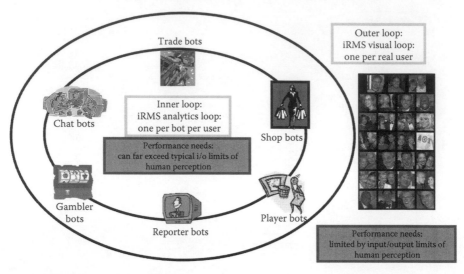

FIGURE 12.10:   Nested RMS instance: virtual world.

photographs, used for mining relevant photos and using them instead to fill the image hole. For example, in our example this would mean finding a photo taken of the building of interest without the car parked in front. Thus, an image processing problem is transformed largely into a data-mining problem, an approach that can only be practical with access to a huge image database.

For another example illustrating algorithmic implications of massive data use, the reader is referred to [4].

12.2.3   Nested RMS

So far we have discussed visual computing loop and analytics loop in isolation. There are certain usage scenarios where these two loops are nested. Consider, for example, a typical virtual world application, such as Second Life. The outer loop in this case is a typical visual computing loop where the end user is interactively synthesizing a virtual world. However, given the multiparty collaborative nature and access to large asset database in this usage, some of the component tasks like retrieving matching digital asset, and whether or not to trade a certain asset, involve search and real-time decision-making as well. One can further imagine a variety of *bots* in an evolved virtual world application mining and analyzing on behalf of a given end user, as shown in Figure 12.10.

12.2.4   Structured Decomposition of RMS Applications

In this section, we propose a structured decomposition of RMS applications. This is an application-algorithm centric decomposition of the application stack. Figure 12.11

Structured decomposition of emerging applications				
Level 0: Services				
Web search	Collaboration	Trading	Gaming	Surveillance
Level 1: Applications				
Semantic search	Artificial intelligence	Portfolio management	Physical simulation	Video mining
Level 2: Mathematical models				
Partitioning based	Neural networks	Lattice-grid models	Mass-spring model	Geometric models
Generative models	Behavioral models	Quadratic optimization	Finite element finite difference	Level sets
Level 3: Mathematical techniques				
SVD	Path planning	Interior-point method	Collision detection	Visual hull
K-means	Stochastic simulation	Simplex	Force solvers	Optical flow
Level 4: Numerical algorithms				
Direct and iterative solvers monte carlo simulations				
Level 5: Numerical primitives and data structures				
Sparse and dense BLAS123 Graphs and matrices Geometry primitives (e.g., triangle, box, convex) Partition structures (e.g., grids, Kd-tree, BVH)				

FIGURE 12.11: Structured decomposition of emerging applications. (Adapted from Chen, Y.-K. et al., *Proc. IEEE*, 96(5), 790, April 2008.)

illustrates six-stage decomposition: *services, applications, mathematical models, mathematical techniques, numerical algorithms,* and *numerical primitives and data structures.* Items listed at each level are not meant to be an exhaustive list, but just an illustrative subset. At the highest level, the majority of these applications are growing into Web-based services. As noted earlier, this makes it easier to constantly evolve the model goodness using the vast temporal and spatial aggregation ability of the Web. Underneath these services, there are classes of related applications. For example, a popular Web search service of modern times is made of applications like text mining, photo search, and video mining. Each one of these applications at next level is composed of a set of analytical or data-driven models of its target entities, such as mass-spring model for cloth simulation used in an online game service. A set of mathematical techniques get deployed for real-time solution of various models. For example, one may use *simplex* method for solving the linear model of a system involving multiple constraints expressed as a system of linear equations. These mathematical techniques in turn rely on a class of numerical algorithms, such as direct or iterative solvers. At the lowest level in this stack lie foundational numerical primitive functions, such as various dense and sparse matrix algebra routines, along with associated data structures like various partitioning structures used in modeling 3D world, such as Kd-tree or bounding volume hierarchies or BVH.

12.3 System Implications

12.3.1 Nature and Source of Underlying Parallelism

There is abundance of parallelism in any RMS application. At the highest level, majority of these applications, as noted earlier, are Web-based service applications, and hence, the multiuser level parallelism opportunity, such as processing thousands of independent queries in a search data center, is most evident. These applications are further distinguished by an abundance of parallelism within a single-user task as well, for example, parallelism within a single trading task for a program trader, or within a single search query, or the same within one fluid simulation task. Data-level parallelism is the most obvious and easiest form of abundant parallelism in the majority of RMS applications. Multimedia datasets can often be spatially or temporally decomposed, processed by a common set of kernels, and effectively integrated at the end. A less obvious source of the intra-task parallelism comes from the growing complexity of the underlying model. Most models discussed in the preceding text are too complex for a direct solution, and instead get solved via a traditional divide-and-conquer approach through decomposition into simpler models. For example, most higher-order estimation tasks can be approximated with multiple lower-order estimation tasks (such as piecewise linear approximation).

Additional source of parallelism during the recombination phase of the subtasks arises from the redundant nature of most decomposition. For example, among the feasible orderings for recomposition, some will have better rate of convergence to the given level of solution quality than others. A sequential algorithm is more dependent on picking the right ordering than a brute-force parallel implementation (when resource feasible). In fact, often the complexity of picking the right ordering of decomposed subtasks on highly resource-constrained sequential machine implementations outweighs the potential gain of parallel decomposition and execution. Therefore, redundant parallel execution may sometimes be preferred. For example, path-finding algorithm implementations on sequential or less-parallel machines often rely on the goodness of sub-path picking heuristics; whereas, the same on a highly parallel machine is less dependent on these, often complex, heuristics.

12.3.1.1 Approximate, Yet Real Time

Most RMS applications have an output that is statistical in nature. Take, for example, a search result, or synthetic reconstruction. There is no *perfect* answer in such cases. Often the quality of the answer improves with more data or more compute, but it can reach the *good-enough* level quite rapidly in many real-life, real-time scenarios. In data-mining literature, such applications are often characterized as *soft* computing applications [5]. Approximate, yet real time, has significant implications on nature and the amount of parallelism. Dependences can be removed exposing more parallelism, or a faster solution, if one is willing to accept an approximate answer. For example, end-biased histogram (part of the so-called iceberg query) is a good example of an approximate query. It answers how often the most frequent item exceeds a certain threshold, as opposed to actual count for the most frequent item. Space complexity of the latter has a provable lower bound of $\Omega(N)$, whereas the end-biased histogram query has a sublinear space complexity. Such reductions in space and time complexities [6] are critical for the underlying streaming (often in-memory) usages for many RMS applications. In general, the statistical nature of most RMS applications lends them to implementations based on randomized algorithm (such as Monte Carlo). As we know, randomized algorithms generally have a higher degree of parallelism, and hence a faster solution, compared to their deterministic counterparts [7].

Furthermore, a significant subset of RMS applications is targeted at digital content creation or audiovisual synthesis. Driven by the needs and limitations of human sensory perceptions, one can make various algorithmic and accuracy trade-offs that would not be permissible in scientific discovery context. For example, a simulation of fluid, aimed at visual fidelity alone, can approximate fluid with a collection of independent particles; whereas, the same for scientific accuracy would mandate a much more complex simulation, in line with the venerated Navier–Stokes equation. These accuracy approximations, when acceptable to specific human perception needs, similar to the previous section, offer new opportunities for parallelization. There are limited and more recent attempts of compile-time optimizations as well for automating the discovery of such performance opportunities in real-time applications, as in the *loop perforation* technique proposed in [8].

12.3.1.2 Curse of Dimensionality and Irregular Access Pattern

Consider a medical imaging context where 2D slices are used for reconstructing the 3D volume of an object, such as the human heart. Progression of heart movements over time adds a fourth dimension. A collection of these volume sequences over different patients or populations can raise the problem dimension to five. The growth of dimension is often much worse for nonphysical objects, such as a transaction record. Thanks to the growing popularity and standardization around XML, high-dimensional objects are quite common. Given this high inherent dimensionality of common models, high-dimensional indexing becomes a challenge for most data access systems. Furthermore, most of the emerging digital data today is unstructured (text, media, blogs, etc.), unlike well-structured, relational databases of the recent past. High-dimensional and unstructured references to such data structures manifest themselves in the form of irregular memory accesses, raising the value of hardware and architectural support for reducing the programming complexity of dealing with this, for example, the gather-scatter feature found in some modern processors [9,10].

12.3.1.3 Parallelism: Both Coarse and Fine Grain

As noted earlier, all RMS applications exhibit a very large degree of parallelism. However, often the large parallelism is fine grain in nature, say at the level of a small kernel, looping over a large stream of data, or a large number of parallel threads of widely varying sizes, including some very small in execution duration. Furthermore, since the subtasks, such as lower-order surfaces, are part of a higher-order model, low overhead sharing of control and data is often crucial for an efficient implementation. Consider, for example, the three levels of parallelism described for parallelizing Cholesky factorization in [11]. There is coarse-grain parallelism at the level of *elimination tree*, a task dependence graph that characterizes the computation and data flow among the *supernodes*. However, the parallelism inside each *supernode* is fine grain in nature. Modern chip-level multiprocessors (CMP) have significantly reduced the traditional overhead associated with fine-grain control and data sharing, as the on-chip processing nodes are only nanoseconds apart from each other, and not milliseconds apart in older HPC systems with comparable computing power. Smelyanskiy [11] demonstrates the potential of very high CMP speedup for a very important, yet traditionally hard-to-parallelize optimization problem of *interior point*, benefiting from the architectural support for both coarse- and fine-grain parallelism.

12.3.1.4 Throughput Computing and Manycore

Abundance of parallelism is the primary distinguishing characteristic of RMS application class. As a result, it is often possible to trade off subtask execution time in favor of improved overall solution time. This is simply an instance of the classic response time versus throughput trade-off in favor of the latter that characterizes *throughput computing*. In other words, at a system level, one can often justify slower single-thread response time in favor of overall throughput performance. RMS applications naturally lend themselves to throughput computing platforms. Fred Pollack [12] observed this during early days of multicore processors. He noted that while single-thread performance typically grows at square root of two for every doubling of transistor budget following venerated Moore's law, for parallel problems with enough coarse-grain parallelism one may be better off using multiple simpler cores in favor of a single big core. Whereas traditional multicore systems strive to deliver idealcase single-thread performance, *manycore* systems, by definition, make a conscious trade-off that reduces single-thread performance by choosing simple, less powerful, individual cores, in favor of increase in overall computational density or throughput of the compute platform.

12.3.1.5 Revisiting Amdahl's Law for Throughput Computing

Let us try to capture the qualitative observations described in the preceding text in quantitative terms. We start with venerated *Amdahl's law*, repeated in the following.

Parallel speedup, $S = ((1 - P) + P/N)^{-1}$, where P is parallel fraction, and N refers to number of processing units. It should be noted that this classic formulation of Amdahl's law implies performance increase to be a monotonic function of N. That is, for a given P, one observes diminishing speedup return with increasing N, but the speedup is always positive. Throughput computing, as defined earlier, violates this aspect of Amdahl's law, as it allows trading off single-thread performance for higher compute density, N. This trade-off is especially relevant in the modern fixed cost (area power) CMP designs. We capture it with the following reformulation: $S' = ((1-P)^*K_N + P/N)^{-1}$. Note the factor, K_N applied to scalar component, implying a slowing down of single-thread performance by a factor that depends on target N. For example, one may be able to offer an $8\times$ increase in compute density from $N = 4$ to $N = 32$, provided we are willing to accept a $4\times$ slowdown on single-thread performance, or $K_{32} = 4$. Under such conditions, performance benefit of higher N is not applicable to all applications as before, rather only to applications with high enough P. Thus, the proposed reformulation captures the high-level trade-off implied by throughput computing, and it no longer has monotonic performance implication for all parallel applications. With a little bit of algebra, one can derive the minimum parallelism, P_{min}, for positive speedup as, $P_{min} = N(K_N - 1)/N(K_N - 1)$. For the illustrative set of parameters here, P must be greater than 0.75 for there to be positive speedup. There have been several recent attempts at revisiting Amdahl's law. The interested reader is referred to [13–15].

12.3.2 Scalability of RMS Applications

Most of us can imagine a large amount of coarse-grain, multiuser parallelism in a large data center context (e.g., thousands of users each trying to launch a Web query, or each trying to simulate a physical object, such as a face, fluid, cloth, etc.). However, it is encouraging to note that there is a considerably high degree of parallelism even within a single-user task, for example, indexing of a text file, or single instance of fluid, face, or cloth simulation. Figure 12.12 shows high-level scalability analysis of a large variety of core compute kernels and applications, each in the context of a single problem instance or a single end user. This performance scalability data is based on cycle-accurate simulation of a manycore, chip-level multiprocessing platform consisting of simpler, throughput-optimized cores, aimed at achieving higher compute density, in line with the observations made in Section 12.3.1.2.

The following observations can be made:

- Quite a few of the applications exhibit large amount of parallelism, and hence near-linear scalability, and nearly all of them show better than 50% resource utilization up to large core counts.

- Primary scalability challenge with many of the applications (e.g., fluid simulation) has less to do with lack of parallelism, and more to do with not being able to data-feed the parallel instances, in a cost-constrained CMP context. This is often referred to as the *feeding the beast challenge*, as the compute density of multicores is expected to increase faster than the external memory bandwidth, rapidly approaching lower than 0.1B/flop from 1B/flop in a single core era.

- Core compute kernels, such as not-too-large-size LU or FFT, can become limited in parallelism as we decompose them too fine. However, higher-level applications have an easier time scaling with core count, due to the higher degree of parallelism more likely to be found in a higher-level app.

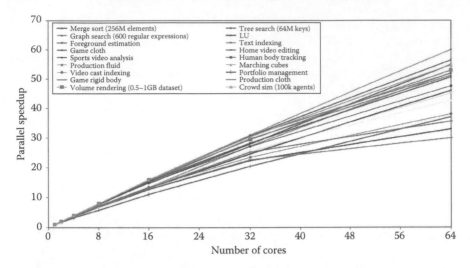

FIGURE 12.12: Scalability of illustrative RMS kernels and applications.

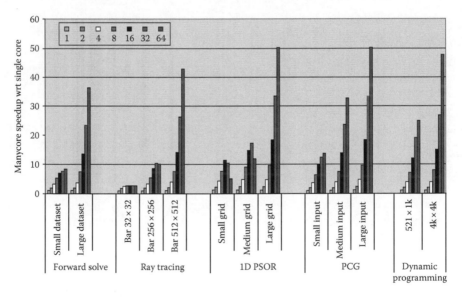

FIGURE 12.13: Scalability improvement with increased problem size.

12.3.2.1 Scalability Implications of Dataset Growth

Scalability data presented in Figure 12.12 is each for a fixed problem size. Scaling the problem size itself can be yet another source for improved scalability. This observation is often referred to as the *Gustafson corollary to Amdahl's law* [16]. This is quantitatively illustrated for some sample cases in Figure 12.13.

12.3.3 Homogenous versus Heterogeneous Decomposition

An important first step in parallelizing an application involves decomposing it into subtasks such that the subtasks are largely independent in their control and data needs. The logical second step involves mapping these tasks onto available compute resources, such that execution time is minimized. Whereas processing nodes on a typical CMP are normally homogenous, this is not necessarily the case for a processing platform with one or more special-purpose accelerators, such as graphics processing units (GPUs) or field programmable gate arrays (FPGAs). A heterogeneous platform, in principle, can offer the most optimal performance. For example, consider the Amdahl's law reformulation discussed in Section 12.3.1.5. In a heterogeneous platform, one can achieve a performance higher than corresponding homogenous multicore/manycore alternatives, by running the scalar portion of the code on a single-thread-optimized integrated core, while running the parallel portion on the large number of throughput-optimized cores. This removes the single-thread performance penalty coefficient, K_N, while still benefitting from the compute density improvement from integration of a large number of high throughput cores.

However, the programming challenges associated with heterogeneous platforms and the difficulty of designing an efficient interface between the various types of compute resources can be nontrivial. The different processing element types may be in a noncoherent memory space. Sharing of data or inter-task communication may only happen under explicit programmer control, and hence additional programming complexity. A decomposition that does not meet the compute, interconnect bandwidth, and capacity constraints of underlying resources can result in significant performance degradation. This challenge is often more acute for heterogeneous systems. Consider, for example, the performance challenge described in [17], where the PPE performance proved to be an overall performance bottleneck for target online gaming application, significantly reducing the overall computational efficiency of the Cell platform. Heterogeneous computing resources, such as a CPU and a GPU, are often connected over an off-chip interconnect, such as PCIe, and invoked through a high overhead driver model [18]. Reference [9] describes a data compression scheme to partly alleviate the interconnect bandwidth limitation. In summary, a parallel decomposition challenge does not end with a parallel algorithm. Often a careful consideration needs to be given to compute resource limitations for achieving performance scaling. For the sake of completeness, we should note that the heterogeneous decomposition problem described thus far gets significantly more complicated when applied to a distributed platform of a variety of clients and servers connected over the Internet, in a peer-to-peer or client-server configuration.

12.4 Conclusion

System-level analysis of important applications leading to improved workload and benchmark proxies is not new. However, the present multicore era has added a dimension of urgency to it. This urgency is especially acute for the compute-intensive subset of applications for the following simple reason: it is very easy for a programmer to get a performance crippling mismatch between the assumed compute model and the host machine model of a modern multicore/manycore, highly threaded processor, than the same in the case of a pre-multicore era, single-core, single-thread processor. In other words, multicore era potentially implies a much higher degree of performance variability than the preceding single-core generation. Coping with this challenge requires a deeper and structured approach to understanding the nature of emerging compute-intensive applications. It is hoped that the RMS taxonomy and the proposed structured framework for application analysis, introduction to the key attributes of the emerging applications and their system implications, has provided the reader the necessary background for delving deeper into this subject. Real-time availability of massive data for a vast majority of tomorrow's computer users, coupled with the rapidly growing compute capabilities of multicore/manycore compute nodes in modern datacenters, offers an unprecedented opportunity for enabling new usages of compute, perhaps making it almost as critical and yet implicit as electricity today.

References

1. Y.-K. Chen, J. Chhugani, P. Dubey, C. J. Hughes, D. Kim, S. Kumar, V. W. Lee, A. D. Nguyen, M. Smelyanskiy. Convergence of recognition, mining, and synthesis workloads and its implications. *Proceedings of the IEEE*, 96(5), 790–807, April 2008.

2. http://www.nature.com/news/2006/061106/full/news061106-6.html

3. J. Hays and A. A. Efros. Scene completion using millions of photographs. In *Siggraph 2007*. Also *Communications of the ACM*, 51(10), 87–94, October 2008.

4. S. Agarwal, N. Snavely, I. Simon, S. M. Seitz, and R. Szeliski. Building Rome in a day. In *International Conference on Computer Vision*, Kyoto, Japan, 2009.

5. S. Mitra, S. K. Pal, and P. Mitra. Data mining in soft computing framework: A survey. *IEEE Transactions on Neural Networks*, 13(1), 3–14, 2002.

6. B. Babcock, S. Babu, M. Datar, R. Motwani, and J. Widom. Models and issues in data stream systems. In *ACM PODS 2002*, Madison, WI, June 3–6, 2002.

7. R. Motwani and P. Raghavan. *Randomized Algorithms*. Cambridge University Press, New York, 1995.

8. S. Misailovic, S. Sidiroglou, H. Hoffmann, and M. Rinard. Quality of service profiling. In *ICSE'10*, Cape Town, South Africa, May 2–8, 2010.

9. M. Smelyanskiy, D. Holmes, J. Chhugani, A. Larson, D. M. Carmean, D. Hanson, P. Dubey, K. Augustine, D. Kim, A. Kyker, V. W. Lee, A. D. Nguyen, L. Seiler, and R. Robb. Mapping high-fidelity volume rendering for medical

imaging to CPU, GPU and many-core architectures. In *IEEE Transactions on Visualization and Computer Graphics*.

10. S. Tzengy, A. Patney, and J. D. Owens. Task management for irregular-parallel workloads on the GPU. In *High Performance Graphics*, Saarbruecken, Germany, June 25–27, 2010.

11. M. Smelyanskiy, V. W Lee, D. Kim, A. D. Nguyen, and P. Dubey. Scaling performance of interior-point method on large-scale chip multiprocessor system. In *Proceedings of the 2007 ACM/IEEE Conference on Supercomputing*, Reno, NV, 2007.

12. Pollack, F. New microarchitecture challenges for the coming generations of CMOS process technologies. IEEE Micro32 Keynote address, 1999.

13. M. D. Hill and M. R. Marty. Amdahl's law in the multicore era. *IEEE Computer*, 41(7), 33–38, July 2008.

14. S. Eyerman and L. Eeckhout. Modeling critical sections in Amdahl's law and its implications for multicore design. In *International Symposium on Computer Architecture (ISCA 2010)*, Saint-Malo, France, June 19–23, 2010.

15. J. M. Paul and B. H. Meyer. Amdahl's law revisited for single chip systems. *International Journal of Parallel Programming*, 35(2), 101–123, April 2007.

16. J. L. Gustafson. Reevaluating Amdahl's law. *Communications of the ACM*, 31(5), 532–533, May 1988.

17. B. D'Amora, A. K. Nanda, K. A. Magerlein, A. Binstock, and B. Yee. High-performance server systems and the next generation of online games. *IBM Systems Journal*, 45(1), 103–118, 2006.

18. L. G. Szafaryn, K. Skadron, and J. J. Saucerman. Experiences accelerating MATLAB systems biology applications. In *Proceedings of the Workshop on Biomedicine in Computing: Systems, Architectures, and Circuits (BiC) 2009*, in conjunction with the *36th IEEE/ACM International Symposium on Computer Architecture (ISCA)*, June 2009.

19. C. Kim, J. Chhugani, N. Satish, E. Sedlar, A. D. Nguyen, T. Kaldewey, V. W. Lee, S. Brandt, P. Dubey. FAST: Fast architecture sensitive tree search on modern CPUs and GPUs. To appear at *2010 ACM SIGMOD/PODS Conference*.

Index

Printed and bound by CPI Group (UK) Ltd, Croydon, CR0 4YY

18/10/2024

01776244-0009